SINTERING OF FUNCTIONAL MATERIALS

Edited by **Igor Shishkovsky**

Sintering of Functional Materials
http://dx.doi.org/10.5772/65530
Edited by Igor Shishkovsky

Contributors

Sutharsini Ubenthiran, Ramesh Singh, Murugathas Thanihaichelvan, Sukjoon Hong, Abdul Shakoor, Penchal Reddy Matli, A.M.A Mohamed, Ismayadi Ismail, Idza Riati Ibrahim, Rodziah Nazlan, Radomir Sokolar, Ben Ayed Foued, Ayadi Ibticem, Raul Valenzuela, Giulia Franceschin, Nancy Flores-Martinez, Gabriela Vazquez-Victorio, Souad Ammar, Walace Matizamhuka

Published by InTech

Janeza Trdine 9, 51000 Rijeka, Croatia

Publishing Process Manager Nina Kalinic

Technical Editor SPi Global

Cover InTech Design team

First published February, 2018

Printed in Croatia

Legal deposit, Croatia: National and University Library in Zagreb

Additional hard copies can be obtained from orders@intechopen.com

Sintering of Functional Materials, Edited by Igor Shishkovsky

p. cm.

Print ISBN 978-953-51-3756-6

Online ISBN 978-953-51-3757-3

For EU product safety concerns:
IN TECH d.o.o., Prolaz Marije Krucifikse Kozulić 3, 51000 Rijeka, Croatia,
info@intechopen.com or visit our website at intechopen.com.

Meet the editor

Igor Shishkovsky is a leading researcher in PN Lebedev Physical Institute (LPI) of the Russian Academy of Sciences (RAS), Samara Branch. His research and teaching focuses on the powder-bed fusion methods and net-shape fabrication of engineering materials via laser sintering techniques in the Russian Federation (RF). Shishkovsky is credited with significant developments in the additive manufacturing technologies. His book, *Laser Synthesis of Functional Mesostructures and 3D Parts,* is the first published and the most cited reference in the field of RF. Shishkovsky has published more than 250 articles, 8 patents, and 7 books in this field.

Contents

Preface

Powder-based materials and treatment technologies rank high in contemporary scientific-technical progress due to their numerous significant technoeconomic qualities. Sintering of such materials allows saving on materials and lowering the cost price of the product, as well as manufacturing complex composite materials with unique combinations of qualities. Materials of record high values of some physic-mechanical and also biochemical characteristics can be obtained owing to structural peculiarities of super dispersed condition.

Sintering of functional materials for innovative perspectives in automotive and aeronautical engineering, space technology, lightweight construction, mechanical engineering, modern design, and many other applications requires established relationship in the materials-process-properties system. Therefore, the industry being interested in understanding theoretical modeling and control over behavior of such powdered materials has promoted the research activities of this manuscript's authors.

The authors of the chapters hope that this book will serve as introduction into the world of sintering of functional materials for the graduate and postgraduate students in materials science and related disciplines.

Prof. Igor Shishkovsky
Lebedev Physical Institute of the Russian Academy of Sciences, Samara Branch,
Samara, Russian Federation

Section 1

Sintering of Functional Materials

Chapter 1

Two-Step Sintering of Ceramics

Ubenthiran Sutharsini,
Murugathas Thanihaichelvan and Ramesh Singh

Additional information is available at the end of the chapter

http://dx.doi.org/10.5772/68083

Abstract

Sintering is a critical phase in the production of ceramic bodies. By controlling the density and microstructure formation, sintering now emerged as a processing technology of ceramic materials. Tailoring the structural, mechanical, electrical, magnetic and optical properties is widening the application of ceramics in various fields. Recently, many advanced sintering methods have reported to fabricate ceramic materials with controlled properties. Two-stage sintering (TSS) is one of the simple and cost-effective methods to obtain near-theoretical density materials with controlled grain growth without adding any dopants. Many recent works have reported the use of TSS as a processing method to fabricate nanoceramics for various applications. With this background, this chapter reviews the advantages of TSS in ceramic preparation based on properties and materials and explores the future directions.

Keywords: two-step sintering, grain growth, ceramic properties, densification, sintering mechanism

1. Introduction

Highly dense ceramics with smaller grain size are widely used in high-performance applications in extreme conditions. Sintering is the responsible step for densification of ceramic bodies, and due to its influence on the properties of the material, sintering is also emerging as a new fabrication method. Controlling the powder size [1], use of sintering additives [2], green body density [3–5] and sintering environment [6, 7] and using new sintering methods such as microwave sintering [8, 9], pressure-assisted sintering [10], spark plasma sintering [11] and field-assisted sintering [3] are used for fabrication of dense and fine-grained ceramics. But these may destroy the unique properties of ceramics [12]. Also the applications of new sintering techniques are limited by low mass productivity, and they are not economically feasible. Two-stage sintering (TSS) is an effective way to achieve fine-grained microstructured ceramics with high densification and relatively low cost. TSS method is successfully applied for all types of ceramics such as structural ceramics, bioceramics, ferrites, piezoelectric ceramics and electrolyte ceramics. Most of the ceramics exhibit controlled or no grain growth in the final stage of sintering and achieved near-theoretical densities. The fine-grained microstructure enhances the mechanical, electrical, magnetic as well as piezoelectric properties of ceramics which widen the applications of ceramics.

TSS consists of heating the samples in two stages. Different sintering profiles were applied in TSS. Generalized diagrams of TSS are shown in **Figure 1(a)** and **(b)**. In the sintering profile 1, the first-step sintering temperature is higher than the second-step sintering temperature, and the first-step holding time is lower than the second-step holding time. Sintering profile 2 is the other way around. The first-step sintering temperature is lower than the second-step sintering temperature, and the first-step holding time is normally higher than the second-step holding time.

In addition, a modified two-step sintering profile is also reported with a cooling step (to room temperature) in between first and second step sintering [2, 13, 14]. Especially, this method used different sintering methods at first and second steps [2, 14].

This chapter mainly focuses on TSS with high first-step sintering temperature. The mechanism of densification with controlled grain growth is explained briefly, and the extended applications of TSS on different ceramics are outlined. Finally the chapter concludes the current trends and challenges of TSS as a fabrication method.

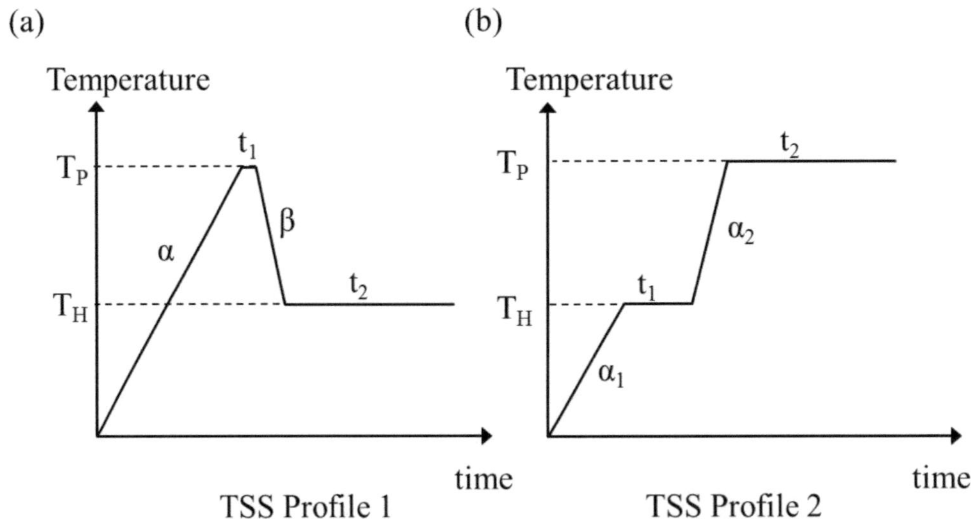

Figure 1. Different sintering profiles used in TSS (a) with high first-step sintering temperature and (b) with a low first-step sintering temperature.

2. TSS with high first-step sintering temperature

TSS with a high first-step sintering temperature is widely used to obtain fully dense ceramics with controlled grain growth due to lower second-step sintering temperature. The sintering profile for TSS with higher first-step sintering temperature is depicted in **Figure 1(a)**. Sintering profile 1 also has few categories based on other sintering parameters:

- Holding time (t_1): The first-step sintering holding time (t_1) is assumed to be zero in few studies, and others hold for a few minutes at temperature T_P.
- Cooling rate (β): Few studies assumed sample rapidly cooled from first sintering temperature (T_P) to second-step sintering temperate (T_H), and others used controlled cooling rates.

The successful TSS method using profile 1 was first introduced for Y_2O_3 by Chen and Wang [15]. In the TSS method, the ceramic samples were heated to a higher temperature to achieve critical density and then immediately cooled to lower temperature and held at that temperature for long holding time to achieve full densification. Density of the sintered sample increased with increasing grain size during the first-step sintering. However, grain growth is controlled at the second stage of sintering, and grain size versus density graph is horizontal at the final stage of the sintering [15]. Similar relationship was observed for other ceramics such as Mg, Nb-doped Y_2O_3 [15], ZnO [16], Ni-Cu-Zn ferrite ceramics, $BaTiO_3$ [17] and $0.89Bi_{0.5}Na_{0.5}TiO_{3}0.06BaTiO_{3-}0.05K_{0.5}Na0.5NbO_3$ lead free antiferroelectric ceramics [18]. However, the most crucial task in this method is to identify suitable sintering parameters such as heating and cooling rate, the first- and second-step sintering temperatures, holding times and sintering atmospheres.

2.1. Sintering mechanism

TSS with a high first-step sintering temperature is widely used to obtain fully dense ceramics with controlled grain growth due to lower second-step sintering temperature. The mechanism for controlled grain growth in TSS with higher first-step sintering temperature was firstly proposed by Chen and Wang [15], for a TSS study on Y_2O_3 ceramics, and it is widely accepted and verified for other ceramics.

The general mechanisms that are responsible for densification during sintering are grain boundary migration and grain boundary diffusion. Grain boundary migration is responsible for the rapid grain growth in the final stage of conventional sintering. The densification of ceramics with grain growth suppression at the second-step sintering can be explained by the absence of grain boundary migration during the second-step sintering as described in **Figure 2**. In conventional single-step sintering (SSS), the grain growth is accelerated due to grain boundary migration and grain boundary diffusion in the final stage of sintering. Rapid cooling before the second stage of sintering freezes the microstructure by immobilizing the triple-point junctions and continues densification will be achieved by grain boundary diffusion.

For a successful TSS profile with higher first-step sintering temperature, few conditions should be achieved at the end of first-step sintering. The sample must be reaching a critical density at

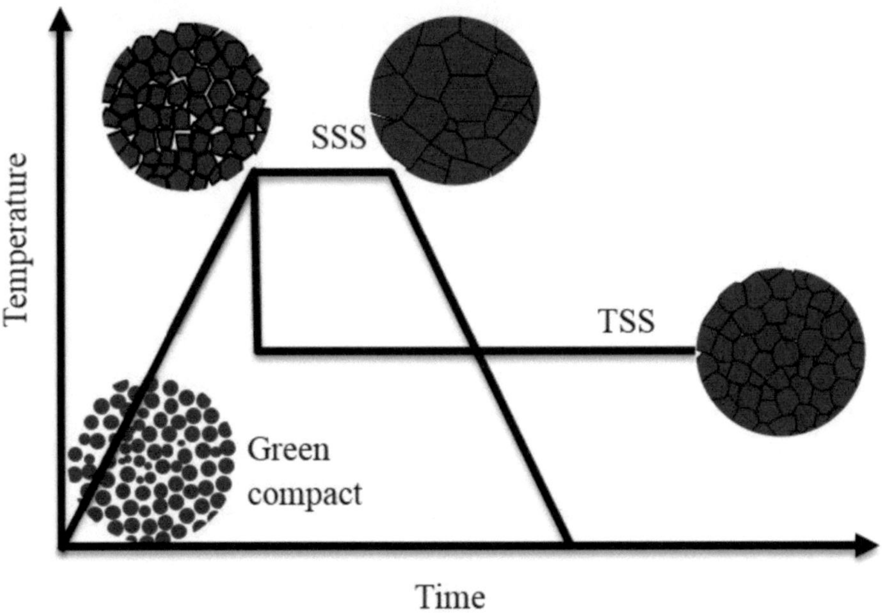

Figure 2. Schematic of densification of ceramics with grain growth during conventional single-step sintering (SSS) and densification without grain growth at lower second-step temperature during TSS with higher first-step temperature.

the end of the first-step sintering so that the densification is active in the final sintering step. This critical density depends on the material. As reported, the critical density to be achieved at the end of the first-step sintering for Y_2O_3 is 75% [15], $BaTiO_3$ is 73% [17] and Ni-Cu-Zn ferrite ceramic is 76% [17]. The critical density is essential to ensure that the pores in the material are subcritical and unstable against shrinkage which can be filled by grain boundary diffusion in the second-step sintering.

A kinetic window will be used to identify the optimum range of the first-step sintering temperature for successful TSS. **Figure 3** illustrates the kinetic window for pure and Mg- and Nb-doped Y_2O_3. The filled squares between the lower and upper limits of the first-step sintering represent the successful second-step sintering. The open squares above the upper limit of the first-step sintering show grain growth during the second step and below the lower limit represents the failed attempts to achieve full densification. It can also be concluded that the kinetic window can be shifted up and down along the first-step temperature axis with the addition of dopants. So far, the kinetic windows have been proposed to Y_2O_3 [15, 19], Mg, Nb-doped Y_2O_3 [15, 19], ZnO [20], Ni-Cu-Zn ferrite [17], $BaTiO_3$ [17] and (1−x) $BiScO_{3-x}$ $PbTiO_3$ (BSPT) [21].

Another important study on exploring the mechanism of grain growth suppression in TSS with higher first-step sintering temperature was reported on ZnO ceramics. The transmission electron microscope (TEM) images of a TSS sample at a first-step sintering temperature of 800°C for 60 s and a second-step sintering temperature of 750°C for 20 h (**Figure 4(a)**) and a conventionally sintered ZnO at 1100°C (**Figure 4(b)**) were examined for the evidences of grain boundary migration. Ten different grain boundary zones were analysed from the TEM

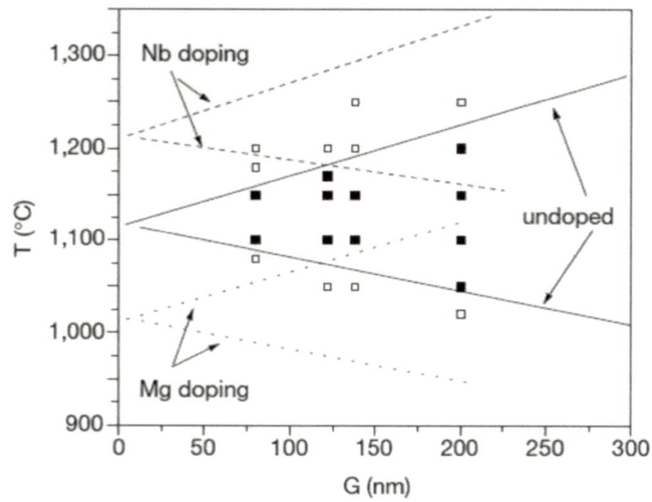

Figure 3. Kinetic window for pure, Mg- and Nb-doped Y_2O_3 (T is the first-step sintering temperature and G is grain size) [15].

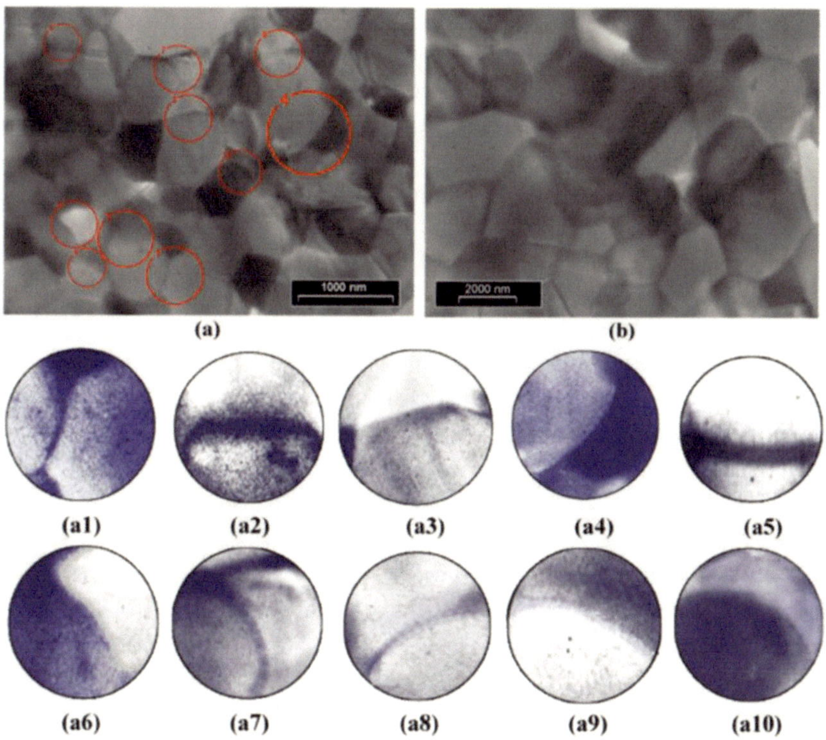

Figure 4. TEM images of the ZnO samples sintered (a) using TSS profile with the first-step sintering temperature of 800°C for 60 s and a second-step sintering temperature of 750°C for 72,000 s, (b) conventionally at 1100°C and (a1)–(a10) magnified images of grain boundaries marked in circles at (a) [22].

image of TSSZnO sample. As can be seen in **Figure 4(a1–a10)**, junctions seem to have pinned the boundaries of growing grains, and the curvature of these boundaries resulted from the mentioned immobilized triple points. However, no similar zones are observed in the TEM images from conventionally sintered samples (**Figure 4(b)**).

2.2. Effect of TSS with high first-step temperature on properties of ceramics

By successfully controlling the grain growth, the TSS with higher first-step sintering temperature is used to fabricate many ceramic materials with enhanced properties that are used for advanced applications [23]. So far TSS is applied for 3 mol% yttria-stabilized tetragonal zirconia polycrystals (3Y-TZP) [1, 4, 5, 8, 24–31], 8 mol% yttria-stabilized zirconia (8YSZ) [3, 5, 9, 24, 32–34], Al_2O_3 [5, 24, 25, 35–43], doped alumina [37, 41], yttrium aluminium garnet (YAG) [44, 45], magnesium aluminium silicate [46], corundum [42, 43], hydroxyapatite [14, 47–50], forsterite (Mg_2SiO_4) [51], TiO_2 [11, 52–54], $SrTiO_3$ [25] $(K_xNa_{1-x})NbO_3$ (KNN) ceramic [12, 55–57], SiC [58–60] and Si3N4 [2] and used for several applications. This section outlined the changes in properties of various ceramics sintered using different TSS profiles.

2.2.1. Sintering of zirconia ceramics

Pure zirconia has three crystallographic structures, monoclinic, tetragonal and cubic. At room temperature monoclinic is stable. In pure form, zirconia has very low appeal for use as engineering ceramic due to reverse transformation on cooling resulting in severe cracking associated with volume expansion (~3%) to the monoclinic phase. In order to overcome this problem, stabilizers such as MgO, CaO and Y_2O_3 are added. Stabilizers reduce the change of chemical free energy and stabilize tetragonal or cubic phase at room temperature.

2.2.1.1. 3Y-TZP ceramics

3Y-TZP is also known as ceramic steel, which exhibits excellent mechanical properties. It is found in many applications, such as cutting tools, grinding media for powders, extrusion dies and biomedical application. Fully dense ceramics with uniform microstructure and fine grain size is essential to stabilize the tetragonal phase as well as to improve the mechanical properties such as hardness and toughness. There are many TSS profiles conducted to control the grain growth of Y-TZP ceramics [1, 4, 5, 8, 24–31, 61]. Only few researchers successfully obtained fully dense ceramics with controlled grain growth (<5 times of powder size) [1, 4, 5, 25, 28, 29, 31, 62]. Among these studies, Binner et al. [31] and others [1, 29] successfully sintered nanostructured zirconia ceramics. Binner et al. [31] achieved fully dense (99% of TD) nanoceramics with the application of hybrid radiant and microwave sintering. However, the second-step grain growth was not entirely suppressed in both radiant and microwave sintering, but the rapid growth observed in conventionally sintered samples was controlled. Application of TSS also enhances the hardness (12–14 GPa), fracture toughness 5–9 $MPam^{1/2}$ and bending strength (900–1100) MPa [61, 62].

Despite the fulfilment of zirconia in a wide variety of application, it suffers from low-temperature degradation in the humid atmosphere around 65–500°C, with hinder biomedical application. Recently Sutharsini et al. [62] sintered fully dense (99% of TD) 3Y-TZP via TSS with average grain size 290 nm. The sintered ceramics exhibited better hydrothermal ageing resistance against the conventionally sintered 3Y-TZP ceramics.

2.2.1.2. *Sintering of 8YSZ*

8YSZ ceramic is a promising candidate as a solid electrolyte for fuel cell application. Maca et al. [5] reported that efficiency of TSS depends on the crystal structure of the ceramics. The authors claimed that cubic structure has higher efficiency in controlling growth than hexagonal and tetragonal structures. Hence, TSS on 8YSZ controls the grain growth compared to 3Y-TZP due to its cubic crystal structure. TSS has been widely applied to 8YSZ to improve mechanical, electrical as well as the gas permeance via controlling the grain growth [3, 9, 24, 32–34]. It enhance the hardness (~13 GPa) [9, 32, 63], fracture toughness (3 $MPam^{1/2}$) [9, 32, 63] and ionic conductivity (0.3 Scm^{-1}) [9]. In addition to the above property, it is necessary to control grain growth to achieve optimum ratio of grain size to electrolyte layer thickness ($0.1 < z < 0.3$) to control gas permeance value for SOFC [64]. TSS effectively controls gas permeance value [26].

Hesabi et al. [9] reported that the TSS is efficient in obtaining fine grain 8YSZ when compared to conventional and microwave sintering with nearly full density. The grain size and conductivity of conventionally sintered sample at 1500°C with a heating rate of 5°C/min, microwave-sintered samples at 1500°C with a heating rate of 50°C/min and TSS with the first-step sintering temperature of 1250°C with no holding time and the second-step sintering temperature of 1050°C with 20 h holding time were compared. The grain sizes were reported as 2.15 μm, 0.9 μm and 295 nm, and the conductivity at 1000°C were 255.4, 322.6 and 398.6 mS/cm for conventional, microwave and two-step sintered samples, respectively.

2.2.2. *Sintering of aluminium-based ceramics*

Alumina-based ceramics are widely used for optical and biomedical applications. It is also used as filler for plastic and cutting inserts. However, the brittle nature of alumina limits the application. Ultrafine grain microstructure is crucial to enhance the mechanical properties such as hardness, wear resistance and strength. TSS successfully applied doped and pure alumina [5, 24, 25, 35–41]. Bodišová et al. [36] and others [5, 25, 35, 37, 39] successfully sintered fully dense alumina (98%) with controlled grain growth (grain growth below five times than that of powder size). Furthermore, the sample sintered with TSS exhibits excellent hardness (18 GPa) and fracture toughness (4 $MPam^{1/2}$) [35].

It is also reported that the room temperature cooling between the first and second steps of sintering also affects the grain growth suppression of corundum abrasives. The TSS with 1250°C first-step temperature and 1350°C with a second-step temperature of 1150°C for 5 h yielded fine-grained corundum abrasives with grain sizes of 65 and 80 nm, respectively. The samples sintered at the same sintering profiles with a room temperature cooling yielded grain sizes of 400 and 560 nm, respectively. The conventional sintering at 1300°C for 2 h resulted to a final grain size of 800 nm [43]. In addition, the two-step sintered corundum abrasive exhibited excellent hardness (22 GPa), fracture toughness (3 $MPam^{1/2}$) and wear resistance (<2 × 10^{-7} mm^3/Nm) [43].

2.2.3. *Sintering of YAG ($Y_3Al_5O_{12}$)*

YAG is a familiar ceramic material for luminescent materials. Presently, single crystalline YAG is applied in lasers pumped by solid-state LEDs, scintillators, and infrared windows. Generally, YAG is sintered for high temperature and long holding time which leads to abnormal grain growth. TSS is an efficient and economic method, which control the abnormal grain growth and improved the transmittance of YAG [44, 45]. YAG sintered with the first-step sintering temperature at 1800°C and the second-step sintering temperature 1550–1600°C revealed more than 40% transmission [44, 45]. Furthermore, neodymium-doped YAG (Nd:YAG) sintered via TSS exhibited excellent transmittance (85%) [65].

2.2.4. *Sintering of hydroxyapatite (HA)*

Hydroxyapatite is a bioceramic that is used as tissue implants due to its excellent biocompatibility. However, low toughness hinders application of artificial bone and teeth implants. Furthermore, the major drawback of HA is that it decomposed into secondary phases (α- or β-tricalcium phosphate). In order to avoid such decomposition, TSS has been applied to HA [14, 47–51, 66, 67]. Feng et al. [14] and others [14, 49, 50, 66, 67] successfully sintered monophase HA without decomposition. Furthermore, TSS improved mechanical properties. Mazaheri et al. [50] achieved highest hardness (7.8 GPa) and fracture toughness (1.9 $MPam^{1/2}$) via TSS.

2.2.5. *Sintering of Ni-Cu-Zn ferrite*

Ni-Cu-Zn ferrite ceramics received special attention due to its low cost, excellent heat and corrosion resistance, high magnetic permeability and low magnetic loss. It is used in many electronic devices such as multilayer capacitor, sensors, antennas and broadband transformers. The electromagnetic properties of Ni-Cu-Zn ferrite are controlled by its microstructure and densification. Wang et al. [17] and Su et al. [68, 69] successfully sintered Ni-Cu-Zn ferrite by using TSS. Wang et al. [17] proposed kinetic window for successful TSS. Ni-Cu-Zn ferrite sintered by using TSS exhibited excellent magnetic properties [68, 69]. Magneto-dielectric materials with matched permeability and permittivity are promising candidates as loading materials to reduce the physical dimensions of low-frequency antennas. Ni-Cu-Zn ferrite sintered via TSS revealed almost equal permeability and permittivity of around 11.8. And the magnetic and dielectric loss tangents were lower than 0.015 in a frequency range from 10 to 100 MHz. These properties make the material useful to the design of miniaturized antennas [69].

2.2.6. *Sintering of Si-based ceramics*

Silicon carbide is widely used for abrasives and refractories due to its high strength, hardness and excellent thermal shock resistance. In conventional single-step sintering, abnormal grain growth is progressed due to its high sintering temperature. Generally, the grain growth during the sintering in the presence of liquid phase is much more significant than that of solid-state sintering. Therefore, it is practically impossible to obtain nanostructured ceramics by conventional single-step liquid-phase sintering. TSS was successfully applied in liquid-phase-sintered SiC ceramics, and a fully dense nanostructured SiC ceramic with a grain size of ~40 nm has been obtained [58, 59] in argon atmosphere. Magnani et al. [60, 70] also successfully sintered doped SiC via TSS with enhanced mechanical properties. The sintered samples exhibited excellent hardness (24 GPa), fracture toughness (3 $MPam^{1/2}$), Young modulus (400 Gpa) and flexural strength (500 MPa).

Similar to SiC, silicon nitride, also non-oxide ceramics, exhibits high hardness strength and thermal shock resistance. It is widely applied to automotive engine wear parts due to its outstanding mechanical properties and wear resistance. Bimodal microstructure of silicon nitride was successfully sintered by using TSS without using β-Si_3N_4 seed crystal [10, 71]. Bimodal microstructure enhances strength and toughness of the ceramic via crack bridging toughness mechanism [72]. Barium aluminosilicate-reinforced silicon nitride sintered via TSS also exhibited higher flexural strength (565 MPa) and fracture toughness (7 $MPam^{1/2}$). The obtained composite exhibits excellent mechanical properties compared to unreinforced barium aluminosilicate matrix [10].

The forsterite (Mg_2SiO_4) ceramic is a new bioceramic with good biocompatibility. Forsterite sintered with the first-step sintering temperature 1300°C and the second-step sintering temperature at 750°C for 15 h revealed high density (98.5%) with grain size 300 nm. Furthermore it exhibited fracture toughness of 3.61 $MPam^{1/2}$. Compared with hydroxyapatite ceramics, forsterite shows a significant improvement in the fracture toughness. Authors suggested that the two-step sintering method can be used to fabricate improved forsterite dense ceramics with desired bioactivity and mechanical properties that might be suitable for hard tissue repair and biomedical applications [51].

2.2.7. Sintering of alkaline niobate-based lead-free piezoelectric ceramics

Environmental friendly lead-free alkaline-based niobate ceramics exhibited excellent piezoelectric properties compared to $Pb(Zr,Ti)O_3$ ceramics. Alkaline-based niobate (KNN) ceramics are successfully sintered by using TSS, and they exhibited excellent dielectric [12, 55, 73], ferroelectric [12, 55, 56, 73] and piezoelectric properties [12, 21, 55, 56]. Furthermore $(K_{0.4425}Na_{0.52}Li_{0.0375})(Nb_{0.8925}Sb_{0.07}Ta_{0.0375})O_3$ exhibited excellent temperature stability over a wide range of temperature, which is attractive for piezoelectric applications [12].

2.2.8. Sintering of zinc oxide

Zinc oxide has been widely applied to electronic and optical devices. Furthermore, alumina-doped ZnO is used as an alternative to indium-doped tin oxide (ITO) as a transparent conductive electrode in photovoltaic devices and displays. Electrical and optical properties of ZnO are mainly influenced by grain size. Grain growth of ZnO was successfully controlled using TSS [16, 20, 22, 74–78]. Zhang et al. [20] and others [16] successfully sintered fully dense

ZnO without grain growth at the final stage of sintering. Furthermore Zhang et al. proposed kinetic window for successful TSS profile. Mazaheri et al. [22] confirmed the triple-point drag mechanism for controlled grain growth at the second-step sintering proposed by Chen and Wang [15] by using TEM image of two-step sintered ZnO compacts. ZnO sintered via TSS exhibited excellent I–V characteristics [16, 74, 76]. ZnO varistors sintered via TSS exhibited higher breakdown field of 6–8 $kVmm^{-1}$ and nonlinear coefficient of over 270 due to fine grain size and high concentration of ZnO-ZnO grain contacts [16].

2.2.9. Sintering of $BaTiO_3$

Barium titanate ($BaTiO_3$) is a polycrystalline piezoelectric ceramic. It is widely used to piezoelectric transducers, sensors and actuators. Many TSS studies have been conducted on $BaTiO_3$ [17, 79–90]. Barium titanate ceramic is widely applied to multilayered ceramic capacitors (MLCC), transducers and pyroelectric detectors due to its dielectric, ferroelectric and piezoelectric properties. Wang et al. [17, 83] successfully sintered fully dense nanostructured ceramic and proposed kinetic window for successful TSS. TSS not only improved the densification and grain growth but also enhanced dielectric and piezoelectric properties [79, 81, 82, 85–88, 91, 92]. TSS samples exhibited excellent piezoelectric constant 519 pN/C and relative permittivity of 6079 [82, 92]. Tian et al. [86] reported that TSS revealed excellent dielectric constant of 2400 at room temperature, low dielectric loss (<1%) and high insulation resistivity of 10^{12} Ωcm, which could be beneficial for multilayer capacitor application.

3. TSS with low first-step sintering temperature

In few TSS studies, samples were initially heated to lower temperatures and then to higher temperature as shown in **Figure 1(b)**. Here the first step is normally a pre-coarsening step that is performed for several purposes including removal of volatile fraction and smoothing the pore channels. This method was successful to prepare fully dense nano-sized pure ZrO_2, fully dense (>98%) alumina [38, 40] and alumina-doped zirconia ($7.5Al_2O_3$.$92.5ZrO_2$) (vol.%) [8].

Sintering of pure zirconia has major drawback due to its reversible tetragonal-to-monoclinic phase transformation associated with shape deformation. Tartaj and Tartaj [93] applied TSS for pure zirconia below the phase transition temperature (<1150°C). In the first step, the compact allowed to achieve 96% of the density at 950°C for 10 h, and then the second-stage sintering temperature increased to 1050°C and achieved fully dense crack-free pure zirconia with grain size less than 200 nm.

Fully dense (99%) alumina-doped zirconia ($7.5Al_2O_3$–$92.5ZrO_2$) (vol.%) is also successfully sintered by using microwave TSS by using lower first-step sintering temperature. Furthermore, microwave-assisted TSS revealed higher density (99%), hardness (13 GPa), fracture toughness (12 $MPam^{1/2}$) and bending strength (750 MPa) than the conventional single-step sintering. Alumina-toughened zirconia is widely applied to dental implant due to its excellent biocompatibility and hardness [8].

Al_2O_3 ceramics are also known as a better translucent with gas-impermeable properties which is suitable for high-pressure lamp envelopes. The high optical transmittance requires special efforts to eliminate any light scattering centres such as residual pores, grain boundaries, secondary phases and rough surfaces in the material. MgO-doped alumina ceramic sintered by TSS with lower first-step sintering exhibited improved the transmittance compared to conventionally sintered sample [40].

4. Conclusion

Despite long holding times, TSS with higher first-step sintering temperature is convenient to achieve fully dense and fine-grained microstructured ceramics with improved properties. The TSS is successfully applied to a range of ceramic materials, and their application is broadened. TSS also helped the emergence of sintering as a fabrication technique. Tailoring the TSS conditions and theoretical studies on TSS mechanisms will make TSS a cost-effective method to fabricate advanced ceramics. TSS can also be studied by using different sintering methods and sintering environments in the first and second steps of sintering.

Author details

Ubenthiran Sutharsini[1]*, Murugathas Thanihaichelvan[1] and Ramesh Singh[2]

*Address all correspondence to: ubsutharsini@gmail.com

1 Department of Physics, Faculty of Science, University of Jaffna, Jaffna, Sri Lanka

2 Department of Mechanical Engineering, Faculty of Engineering, University of Malaya, Kuala Lumpur, Malaysia

References

[1] Suárez G, Sakka Y, Suzuki TS, Uchikoshi T, Zhu X, Aglietti EF. Effect of starting powders on the sintering of nanostructured ZrO_2 ceramics by colloidal processing. Science and Technology of Advanced Materials. 2009;**10**:025004(1-8). DOI: 10.1088/1468-6996/10/2/025004/meta

[2] Kim HD, Han BD, Park DS, Lee BT, Becher PF. Novel two-step sintering process to obtain a bimodal microstructure in silicon nitride. Journal of the American Ceramic Society. 2002;**85**:245-252. DOI: 10.1111/j.1151-2916.2002.tb00073.x

[3] Schwarz S, Guillon O. Two step sintering of cubic yttria stabilized zirconia using Field Assisted Sintering Technique/Spark Plasma Sintering. Journal of the European Ceramic Society. 2013;**33**:637-641. DOI: 10.1016/j.jeurceramsoc.2012.10.002

[4] Xiong Y, Hu J, Shen Z. Dynamic pore coalescence in nanoceramic consolidated by two-step sintering procedure. Journal of the European Ceramic Society. 2013;**33**:2087-2092. DOI: 10.1016/j.jeurceramsoc.2013.03.015

[5] Maca K, Pouchly V, Zalud P. Two-step sintering of oxide ceramics with various crystal structures. Journal of the European Ceramic Society. 2010;**30**:583-589. DOI: 10.1016/j.jeurceramsoc.2009.06.008

[6] Ubenthiran S, Ramesh S, Tan CY, Teng WD. Oxygen vacancy comparisons for 3Y- TZP sintered in air and argon gas atmosphere. Applied Mechanics and Materials. 2013;**372**: 173-176. DOI: 10.4028/www.scientific.net/AMM.372.173

[7] Sutharsini U, Ramesh S, Purbolaksono J, Tan CY, Teng WD, Amiriyan M. Low-temperature degradation and defect relationship in yttria-tetragonal zirconia polycrystal ceramic. Materials Research Innovations. 2014;**18**:S6-131-S136-134. DOI: 10.1179/1432891714Z.000000000943

[8] Ai Y, Xie X, He W, Liang B, Fan Y. Microstructure and properties of Al_2O_3 (n)/ZrO_2 dental ceramics prepared by two-step microwave sintering. Materials & Design. 2015;**65**: 1021-1027. DOI: 10.1016/j.matdes.2014.10.054

[9] Hesabi ZR, Mazaheri M, Ebadzadeh T. Enhanced electrical conductivity of ultrafine-grained $8Y_2O_3$ stabilized ZrO_2 produced by two-step sintering technique. Journal of Alloys and Compounds. 2010;**494**:362-365. DOI: 10.1016/j.jallcom.2010.01.046

[10] Ye F, Liu L, Zhang J, Iwasa M, Su CL. Synthesis of silicon nitride-barium aluminosilicate self-reinforced ceramic composite by a two-step pressureless sintering. Composites Science and Technology. 2005;**65**:2233-2239. DOI: 10.1016/j.compscitech.2005.04.015

[11] Mazaheri M, Zahedi AM, Haghighatzadeh M, Sadrnezhaad SK. Sintering of titania nanoceramic: Densification and grain growth. Ceramics International. 2009;**35**:685-691. DOI: 10.1016/j.ceramint.2008.02.005

[12] Pang X, Qiu J, Zhu K, Du J. (K, Na)NbO_3 based lead free piezoelectric ceramics manufactured by two step sintering. Ceramics International. 2012;**38**:2521-2527. DOI: 10.1016/j.ceramint.2011.11.022

[13] Khosroshahi HR, Ikeda H, Yamada K, Saito N, Kaneko K, Hayashi K, Nakashima K. Effect of cation doping on mechanical properties of yttria prepared by an optimized two-step sintering process. Journal of the American Ceramic Society. 2012;**95**:3263-3269. DOI: 10.1111/j.1551-2916.2012.05379.x

[14] Feng P, Niu M, Gao C, Peng S, Shuai C. A novel two-step sintering for nano-hydroxyapatite scaffolds for bone tissue engineering. Scientific Reports. 2014;**4**:5599. DOI: 10.1038/srep05599

[15] Chen IW, Wang XH. Sintering dense nanocrystalline ceramics without final-stage grain growth. Nature. 2000;**404**:168-171. DOI: 10.1038/35004548

[16] Durán P, Tartaj J, Moure C. Fully dense, fine-grained, doped zinc oxide varistors with improved nonlinear properties by thermal processing optimization. Journal of the American Ceramic Society. 2003;**86**:1326-1329. DOI: 10.1111/j.1151-2916.2003.tb03470.x

[17] Wang XH, Deng XY, Bai HL, Zhou H, Qu WG, Li LT, Chen IW. Two-step sintering of ceramics with constant grain-size, II: $BaTiO_3$ and Ni–Cu–Zn Ferrite. Journal of the American Ceramic Society. 2006;**89**:438-443. DOI: 10.1111/j.1551-2916.2005.00728.x

[18] Ding J, Liu Y, Lu Y, Qian H, Gao H, Chen H, Ma C. Enhanced energy-storage properties of $0.89Bi_{0.5}Na_{0.5}TiO_{3\text{-}0}.06BaTiO_{3\text{-}0}.05K_{0.5}Na_{0.5}NbO_3$ lead-free anti-ferroelectric ceramics by two-step sintering method. Materials Letters. 2014;**114**:107-110. DOI: 10.1016/j.matlet.2013.09.103

[19] Wang XH, Chen PL, Chen IW. Two-step sintering of ceramics with constant grain-size, I. Y_2O_3. Journal of the American Ceramic society. 2006;**89**:431-437. DOI: 10.1111/j.1551-2916.2005.00763

[20] Zhang Y, Tan R, Yang Y, Zhang X, Wang W, Cui P, Song W. Two-step sintering of pristine and aluminum-doped zinc oxide ceramics. International Journal of Applied Ceramic Technology. 2012;**9**:960-967. DOI: 10.1111/j.1744-7402.2011.02702.x

[21] Zou T, Wang X, Zhao W, Li L. Preparation and properties of fine-grain (1−*x*) $BiScO_{3-x}$ $PbTiO_3$ ceramics by two-step sintering. Journal of the American Ceramic Society. 2008;**91**:121-126. DOI: 10.1111/j.1551-2916.2007.01903.x

[22] Mazaheri M, Zahedi AM, Sadrnezhaad SK. Two-step sintering of nanocrystalline ZnO compacts: effect of temperature on densification and grain growth. Journal of the American Ceramic Society. 2008;**91**:56-63. DOI: 10.1111/j.1551-2916.2007.02029.x

[23] He R, Zhang R, Pei Y, Fang D. Two-step hot pressing of bimodal micron/nano-ZrB_2 ceramic with improved mechanical properties and thermal shock resistance. International Journal of Refractory Metals and Hard Materials. 2014;**46**:65-70. DOI: 10.1016/j.ijrmhm.2014.05.016

[24] Pouchly V, Maca K, Shen Z. Two-stage master sintering curve applied to two-step sintering of oxide ceramics. Journal of the European Ceramic Society. 2013;**33**:2275-2283. DOI: 10.1016/j.jeurceramsoc.2013.01.020

[25] Maca K, Pouchly V, Shen Z. Two-step sintering and spark plasma sintering of Al_2O_3, ZrO_2 and $SrTiO_3$ ceramics. Integrated Ferroelectrics. 2008;**99**:114-124. DOI: 10.1080/10584580802107841

[26] Wright GJ, Yeomans JA. Constrained sintering of yttria-stabilized zirconia electrolytes: The influence of two-step sintering profiles on microstructure and gas permeance. International Journal of Applied Ceramic Technology. 2008;**5**:589-596. DOI: 10.1111/j.1744-7402.2008.02263.x

[27] Wright GJ, Yeomans JA. Three-step sintering of constrained yttria stabilised zirconia layers and its effect on microstructure and gas permeance. Journal of the European Ceramic Society. 2009;**29**:1933-1938. DOI: 10.1016/j.jeurceramsoc.2008.12.013

[28] Mazaheri M, Simchi A, Golestani-Fard F. Densification and grain growth of nanocrystalline 3Y-TZP during two-step sintering. Journal of the European Ceramic Society. 2008;**28**:2933-2939. DOI: 10.1016/j.jeurceramsoc.2008.04.030

[29] Lourenço MA, Cunto GG, Figueiredo FM, Frade JR. Model of two-step sintering conditions for yttria-substituted zirconia powders. Materials Chemistry and Physics. 2011;**126**:262-271. DOI: 10.1016/j.matchemphys.2010.11.028

[30] Suarez G, Sakka Y, Suzuki TS, Uchikoshi T, Aglietti EF. Effect of bead-milling treatment on the dispersion of tetragonal zirconia nanopowder and improvements of two-step sintering. Journal of the Ceramic Society of Japan. 2009;**117**:470-474. DOI: 10.2109/jcersj2.117.470

[31] Binner J, Annapoorani K, Paul A, Santacruz I, Vaidhyanathan B. Dense nanostructured zirconia by two stage conventional/hybrid microwave sintering. Journal of the European Ceramic Society. 2008;**28**:973-977. DOI: 10.1016/j.jeurceramsoc.2007.09.002

[32] Mazaheri M, Valefi M, Hesabi ZR, Sadrnezhaad S. Two-step sintering of nanocrystalline $8Y_2O_3$ stabilized ZrO_2 synthesized by glycine nitrate process. Ceramics International. 2009;**35**:13-20. DOI: 10.1016/j.jeurceramsoc.2007.09.002

[33] Rajeswari K, Hareesh U, Subasri R, Chakravarty D, Johnson R. Comparative evaluation of spark plasma (SPS), microwave (MWS), two stage sintering (TSS) and conventional sintering (CRH) on the densification and microstructural evolution of fully stabilized zirconia ceramics. Science of Sintering. 2010;**42**:259-267. DOI: 10.2298/SOS1003259R

[34] Laberty-Robert C, Ansart F, Deloget C, Gaudon M, Rousset A. Dense yttria stabilized zirconia: sintering and microstructure. Ceramics International. 2003;**29**:151-158. DOI: 10.1016/S0272-8842(02)00099-8

[35] Razavi Hesabi Z, Haghighatzadeh M, Mazaheri M, Galusek D, Sadrnezhaad S. Suppression of grain growth in sub-micrometer alumina via two-step sintering method. Journal of the European Ceramic Society. 2009;**29**:1371-1377. DOI: 10.1016/j.jeurceramsoc.2008.08.027

[36] Bodišová K, Šajgalík P, Galusek D, Švančárek P. Two-stage sintering of alumina with submicrometer grain Size. Journal of the American Ceramic Society. 2007;**90**:330-332. DOI: 10.1111/j.1551-2916.2006.01408.x

[37] Galusek D, Ghillányová K, Sedláček J, Kozánková J, Šajgalík P. The influence of additives on microstructure of sub-micron alumina ceramics prepared by two-stage sintering. Journal of the European Ceramic Society. 2012;**32**:1965-1970. DOI: 10.1016/j.jeurceramsoc.2011.11.038

[38] Lin FJT, de Jonghe LC, Rahaman MN. Microstructure refinement of sintered alumina by a two-step sintering technique. Journal of the American Ceramic Society. 1997;**80**:2269-2277. DOI: 10.1111/j.1151-2916.1997.tb03117.x

[39] Michálková M, Ghillányová K, Galusek D. Standard and two-stage sintering of a submicrometer alumina powder. In: RK Bordia, EAOlevsky. Advances in Sintering Science and Technology, John Wiley & Sons, Inc.; Hoboken, NJ, USA; 2010. 421 p. DOI: 10.1002/9780470599730.ch40

[40] Kim DS, Lee JH, Sung RJ, Kim SW, Kim HS, Park JS. Improvement of translucency in Al_2O_3 ceramics by two-step sintering technique. Journal of the European Ceramic Society. 2007;**27**:3629-3632. DOI: 10.1016/j.jeurceramsoc.2007.02.002

[41] Wang CJ, Huang CY, Wu YC. Two-step sintering of fine alumina–zirconia ceramics. Ceramics International. 2009;**35**:1467-1472. DOI: 10.1016/j.ceramint.2008.08.001

[42] Sairam K, Sonber J, Murthy TC, Sahu A, Bedse R, Chakravartty J. Pressureless sintering of chromium diboride using spark plasma sintering facility. International Journal of Refractory Metals and Hard Materials. 2016;**58**:165-171. DOI: 10.1016/j.ijrmhm.2016.05.002

[43] Li Z, Li Z, Zhang A, Zhu Y. Synthesis and two-step sintering behavior of sol–gel derived nanocrystalline corundum abrasives. Journal of the European Ceramic Society. 2009;**29**:1337-1345. DOI: 10.1016/j.jeurceramsoc.2008.09.004

[44] Chen ZH, Li JT, Xu JJ, Hu ZG. Fabrication of YAG transparent ceramics by two-step sintering. Ceramics International. 2008;**34**:1709-1712. DOI: 10.1016/j.ceramint.2007.05.015

[45] Li X, Zheng B, Odoom-Wubah T, Huang J. Co-precipitation synthesis and two-step sintering of YAG powders for transparent ceramics. Ceramics International. 2013;**39**:7983-7988. DOI: 10.1016/j.ceramint.2013.03.064

[46] Liu J, Lv X, Li J, Jiang L. Pressureless sintered magnesium aluminate spinel with enhanced mechanical properties obtained by the two-step sintering method. Journal of Alloys and Compounds. 2016;**680**:133-138. DOI: 10.1016/j.jallcom.2016.04.192

[47] Lukić M, Stojanović Z, Škapin SD, Maček-Kržmanc M, Mitrić M, Marković S, Uskoković D. Dense fine-grained biphasic calcium phosphate (BCP) bioceramics designed by two-step sintering. Journal of the European Ceramic Society. 2011;**31**:19-27. DOI: 10.1016/j.jeurceramsoc.2010.09.006

[48] Panyata S, Eitssayeam S, Rujijanagul G, Tunkasiri T, Pengpat K. Property development of hydroxyapatite ceramics by two-step sintering. In: Advanced Materials Research, Trans Tech Publ; Zurich, Switzerland; 2012, p. 190-193. DOI: 10.4028/www.scientific.net/AMR.506.190

[49] Lin K, Chen L, Chang J. Fabrication of dense hydroxyapatite nanobioceramics with enhanced mechanical properties via two-step sintering process. International Journal of Applied Ceramic Technology. 2012;**9**:479-485. DOI: 10.1111/j.1744-7402.2011.02654.x

[50] Mazaheri M, Haghighatzadeh M, Zahedi AM, Sarnezhaad SK. Effect of a novel sintering process on mechanical properties of hydroxyapatite ceramics. Journal of Alloys and Compounds. 2009;**471**:180-184. DOI: 10.1016/j.jallcom.2008.03.066

[51] Fathi MH, Kharaziha M. Two-step sintering of dense, nanostructural forsterite. Materials Letters. 2009;**63**:1455-1458. DOI: 10.1016/j.matlet.2009.03.040

[52] Mazaheri M, Razavi Hesabi Z, Sadrnezhaad SK. Two-step sintering of titania nanoceramics assisted by anatase-to-rutile phase transformation. Scripta Materialia. 2008;**59**: 139-142. DOI: 10.1016/j.scriptamat.2008.02.041

[53] Li D, Chen SO, Shao WQ, Wang DC, Li YH, Long YZ, Liu ZW, Ringer SP. Preparation of dense nanostructured titania ceramic using two step sintering. Materials Technology. 2010;**25**:42-44. DOI: 10.1179/175355509X12608871787997

[54] Li BR, Liu DY, Liu JJ, Hou SX, Yang ZW. Two-step sintering assisted consolidation of bulk titania nano-ceramics by spark plasma sintering. Ceramics International. 2012;**38**:3693-3699. DOI: 10.1016/j.ceramint.2012.01.011

[55] Fang J, Wang X, Tian Z, Zhong C, Li L, Zuo R. Two-step sintering: An approach to broaden the sintering temperature range of alkaline niobate-based lead-free piezoceramics. Journal of the American Ceramic Society. 2010;**93**:3552-3555. DOI: 10.1111/j.1551-2916.2010.04085.x

[56] Hao J, Bai W, Shen B, Zhai J. Improved piezoelectric properties of $(K_xNa_{1-x})0.94Li_{0.06}NbO_3$ lead-free ceramics fabricated by combining two-step sintering. Journal of Alloys and Compounds. 2012;**534**:13-19. DOI: 10.1016/j.jallcom.2012.04.033

[57] Lartcumfu N, Kruea-In C, Tawichai N, Rujijanagul G. Fabrication of sodium potassium niobate ceramics by two step sintering assisted molten salts synthesis. Ferroelectrics. 2013;**456**:14-20. DOI: 10.1080/00150193.2013.846171

[58] Lee YI, Kim YW, Mitomo M. Effect of processing on densification of nanostructured SiC ceramics fabricated by two-step sintering. Journal of Material Science. 2004;**39**:3801-3803. DOI: 10.1023/B:JMSC.0000030743.62306.70

[59] Lee YI, Kim YW, Mitomo M, Kim DY. Fabrication of dense nanostructured silicon carbide ceramics through two-step sintering. Journal of the American Ceramic Society. 2003;**86**:1803-1805. DOI: 10.1111/j.1151-2916.2003.tb03560.x

[60] Magnani G, Sico G, Brentari A. Two-Step Pressureless Sintering of Silicon Carbide-Based Materials; In: Pietro Vincenzini. Advances in Science and Technology, Trans Tech Publ; Zurich, Switzerland; 2014, p. 70-75. DOI:10.4028/www.scientific.net/AST.89.70

[61] Chao M, Ning L, Zhikai W, Jing T, Jiazhen Y. Influence on microstructure of dental zirconia ceramics prepared by two-step sintering. West China Journal of Stomatology. 2013;**31**, 496-499. DOI: 10.7518/hxkq.2013.05.013

[62] Sutharsini U, Ramesh S, Wong YH, Misran H, Yusuf F, Tan CY, Purbolaksono J, Teng WD. Effect of sintering holding time on low-temperature degradation of yttria stabilised zirconia ceramics. Materials Research Innovations. 2014;**18**:S6-408-S406-411. DOI: 10.117 9/1432891714Z.000000000988

[63] Mazaheri M, Zahedi A, Hejazi M. Processing of nanocrystalline 8 mol% yttria-stabilized zirconia by conventional, microwave-assisted and two-step sintering. Materials Science and Engineering. 2008;**492**:261-267. DOI: 10.1016/j.msea.2008.03.023

[64] Wright GJ, Yeomans JA. The influence of screen-printing parameters on the microstructure and gas permeance of a zirconia electrolyte. Journal of the European Ceramic Society. 2008;**28**:779-785. DOI: 10.1016/j.jeurceramsoc.2007.09.027

[65] Li J, Chen Q, Feng G, Wu W, Xiao D, Zhu J. Optical properties of the polycrystalline transparent Nd: YAG ceramics prepared by two-step sintering. Ceramics International. 2012;**38**:S649-S652. DOI: 10.1016/j.ceramint.2011.05.127

[66] Chu KT, Ou SF, Chen SY, Chiou SY, Chou HH, Ou KL. Research of phase transformation induced biodegradable properties on hydroxyapatite and tricalcium phosphate based bioceramic. Ceramics International. 2013;**39**:1455-1462. DOI: 10.1016/j.ceramint.2012.07.089

[67] Veljovic D, Palcevskis E, Zalite I, Petrovic R, Janackovic D. Two-step microwave sintering—A promising technique for the processing of nanostructured bioceramics. Materials Letters. 2013;**93**:251-253. DOI: 10.1016/j.matlet.2012.11.095

[68] Su H, Tang X, Zhang H, Zhong Z, Shen J. Sintering dense Ni-Zn ferrite by two-step sintering process. Journal of Applied Physics. 2011;**109**:07A501. DOI: 10.1063/1.3535418

[69] Su H, Tang X, Zhang H, Jing Y, Bai F, Zhong Z. Low-loss Ni-Cu-Zn ferrite with matching permeability and permittivity by two-step sintering process. Journal of Applied Physics. 2013;**113**:17B301. DOI: 10.1063/1.4793508

[70] Magnani G, Brentari A, Burresi E, Raiteri G. Pressureless sintered silicon carbide with enhanced mechanical properties obtained by the two-step sintering method. Ceramics International. 2014;**40**:1759-1763. DOI: 10.1016/j.ceramint.2013.07.075

[71] Kim HD, Park YJ, Han BD, Park MW, Bae WT, Kim YW, Lin HT, Becher PF. Fabrication of dense bulk nano-Si_3N_4 ceramics without secondary crystalline phase. Scripta materialia. 2006;**54**:615-619. DOI: 10.1016/j.scriptamat.2005.10.032

[72] Becher PF. Microstructural design of toughened ceramics. Journal of the American Ceramic Society. 1991;**74**:255-269. DOI: 10.1111/j.1151-2916.1991.tb06872.x

[73] Kato K, Kakimoto K, Hatano K, Kobayashi K, Doshida Y. Lead-free Li-modified (Na, K)NbO_3 piezoelectric ceramics fabricated by two-step sintering method. Journal of the Ceramic Society of Japan. 2014;**122**:460-463. DOI: 10.2109/jcersj2.122.460

[74] Pillai SC, Kelly JM, McCormack DE, Ramesh R. Effect of step sintering on breakdown voltage of varistors prepared from nanomaterials by sol gel route. Advances in Applied Ceramics. 2006;**105**:158-160. DOI: 10.1179/174367606X110189

[75] Anas S, Mangalaraja R, Poothayal M, Shukla SK, Ananthakumar S. Direct synthesis of varistor-grade doped nanocrystalline ZnO and its densification through a step-sintering technique. Acta Materialia. 2007;**55**:5792-5801. DOI: 10.1016/j.actamat.2007.06.047

[76] Shahraki M, Shojaee SA, Sani MA F, Nemati A, Safaee I. Two-step sintering of ZnO varistors. Solid State Ionics. 2011;**190**:99-105. DOI: 10.1016/j.ssi.2010.06.026

[77] Chen B, Xia Z, Lu K. Understanding sintering characteristics of ZnO nanoparticles by FIB-SEM three-dimensional analysis. Journal of the European Ceramic Society. 2013;**33**:2499-2507. DOI: 10.1016/j.jeurceramsoc.2013.04.026

[78] Wu MW. Two-step sintering of aluminum-doped zinc oxide sputtering target by using a submicrometer zinc oxide powder. Ceramics International. 2012;**38**:6229-6234. DOI: 10.1016/j.ceramint.2012.04.076

[79] Takuya H, Kayo T, Jianyong L, Takeshi K, Hirofumi K, Takaaki T. Domain size effect on dielectric properties of barium titanate ceramics. Japanese Journal of Applied Physics. 2008;**47**:7607-7611. DOI: 10.1143/JJAP.47.7607/meta

[80] Hoshina T, Kigoshi Y, Furuta T, Takeda H, Tsurumi T. Shrinkage behaviors and sintering mechanism of $BaTiO_3$ ceramics in two-step sintering. Japanese Journal of Applied Physics. 2011;**50**:09NC07-(1-4). DOI: 10.1143/JJAP.50.09NC07

[81] Xu T, Wang CA. Effect of two-step sintering on micro-honeycomb $BaTiO_3$ ceramics prepared by freeze-casting process. Journal of the European Ceramic Society. 2016;**36**:2647-2652. DOI: 10.1016/j.jeurceramsoc.2016.03.032

[82] Huan Y, Wang X, Fang J, Li L. Grain size effects on piezoelectric properties and domain structure of $BaTiO_3$ ceramics prepared by two-step sintering. Journal of the American Ceramic Society. 2013;**96**:3369-3371. DOI: 10.1111/jace.12601

[83] Wang XH, Deng XY, Zhou H, Li LT, Chen IW. Bulk dense nanocrystalline $BaTiO_3$ ceramics prepared by novel pressureless two-step sintering method. Journal of Electroceramics. 2008;**21**:230-233. DOI: 10.1007/s10832-007-9143-1

[84] Moon SM, Wang X, Cho NH. Nanostructural and physical features of $BaTiO_3$ ceramics prepared by two-step sintering. Journal of the Ceramic Society of Japan. 2009;**117**:729-731. DOI: 10.2109/jcersj2.117.729

[85] Tian Z, Wang X, Lee S, Hur KH, Li L. Microstructure evolution and dielectric properties of ultrafine grained $BaTiO_3$-based ceramics by two-step sintering. Journal of the American Ceramic Society. 2011;**94**:1119-1124. DOI: 10.1111/j.1551-2916.2010.04234.x

[86] Tian Z, Wang X, Shu L, Wang T, Song TH, Gui Z, Li L. Preparation of nano $BaTiO_3$ based ceramics for multilayer ceramic capacitor application by chemical coating method. Journal of the American Ceramic Society. 2009;**92**:830-833. DOI: 10.1111/j.1551-2916.2009.02979.x

[87] Karaki T, Yan K, Adachi M. Barium titanate piezoelectric ceramics manufactured by two-step sintering. Japanese Journal of Applied Physics. 2007;**46**:7035-7038. DOI: 10.1143/JJAP.46.7035

[88] Polotai A, Breece K, Dickey E, Randall C, Ragulya A. A Novel approach to sintering nanocrystalline barium titanate ceramics. Journal of the American Ceramic Society. 2005;**88**:3008-3012. DOI: 10.1111/j.1551-2916.2005.00552.x

[89] Kim HT, Han YH. Sintering of nanocrystalline $BaTiO_3$. Ceramics International. 2004;**30**:1719-1723. DOI: 10.1016/j.ceramint.2003.12.141

[90] Chaisan W, Yimnirun R, Ananta S. Two-stage sintering of barium titanate ceramic and resulting characteristics. Ferroelectrics. 2007;**346**:84-92. DOI: 10.1080/00150190601180380

[91] Tomoaki K, Kang Y, Toshiyuki M, Masatoshi A. Lead-free piezoelectric ceramics with large dielectric and piezoelectric constants manufactured from $BaTiO_3$ nano-powder. Japanese Journal of Applied Physics. 2007;**46**:L97-L98. DOI: 10.1143/JJAP.46.L97/meta

[92] Huan Y, Wang X, Fang J, Li L. Grain size effect on piezoelectric and ferroelectric properties of $BaTiO_3$ ceramics. Journal of the European Ceramic Society. 2014;**34**:1445-1448. DOI: 10.1016/j.jeurceramsoc.2013.11.030

[93] Tartaj J, Tartaj P. Two-stage sintering of nanosize pure zirconia. Journal of the American Ceramic Society. 2009;**92**:S103-S106. DOI: 10.1111/j.1551-2916.2008.02723.x

Chapter 2

Development of Metal Matrix Composites Using Microwave Sintering Technique

Penchal Reddy Matli, Rana Abdul Shakoor and Adel Mohamed Amer Mohamed

Additional information is available at the end of the chapter

http://dx.doi.org/ 10.5772/68081

Abstract

In this book chapter, aluminum (Al)-based metal matrix composites (AMMCs) with various reinforcing ceramic particles, such as SiC, Si_3N_4, and Al_2O_3, were produced by microwave sintering and subsequent hot extrusion processes. The role of various nano/micro-sized reinforcements in altering the structural, mechanical, and thermal properties of the microwave-extruded composites was systematically studied. The X-ray diffraction (XRD) patterns indicated that the main components were Al, SiC, Si_3N_4, and Al_2O_3 for the studied Al-SiC, Al-Si_3N_4, and Al-Al_2O_3 composites, respectively. Scanning electron microscopy (SEM) and energy dispersive spectroscopy (EDS) elemental mapping confirm the homogeneous distribution of reinforcing particles in the Al matrix. Mechanistic studies revealed that the Al-Si_3N_4 metal matrix composite exhibited superior hardness, ultimate compression/tensile strength, and Young's modulus, while having a lower coefficient of thermal expansion compared to other studied Al composites. Findings presented are expected to pave the way to design, develop, and synthesize other aluminum-based metal matrix composites for automotive and industrial applications.

Keywords: Al matrix composites, ceramic reinforcements, microwave sintering, hot extrusion, mechanical properties, thermal properties, fracture behavior

1. Introduction

Microwaves occupy the portion of electromagnetic radiation spectrum between 300 MHz and 300 GHz with wavelengths ranging from 1 mm to 1 m in free space. Although the frequencies available for processing of materials are 24.124 GHz, 5.8 GHz, 2.45 GHz, and 915 MHz, generally it is carried out at 915 MHz and 2.45 GHz. Usually, 9.15 and 2.45 GHz are commonly used in industrial equipment [1, 2]. Microwave cavities are of two types: single-mode resonance cavities and multimode resonance cavities. Single-mode cavities are specially designed and generally used for industrial applications. The domestic microwave ovens are multimode cavities in which multiple plane waves impinge on the load (material to be heated) from different directions. The microwave radiation and heating were applied to the fabrication of ceramic materials till the 1990s [3–5].

1.1. Characteristics and merits of microwave sintering process

Microwave sintering has attained global acceptance because of superior benefits over the traditional sintering techniques. The characteristic (**Figure 1**) of microwave heating is basically different from conventional heating [6]. For conventional heating, an external heating element is used for heat generation and then it transferred to the test materials through convention, conduction, and radiation. In microwave heating, the heat is generated internally within the test sample by rapid oscillation of dipoles at microwave frequencies [7], instead of diffusion from external sources and hence the heating is from the core of the sample to outside. Heating is very rapid and volumetric due to energy conversion rather than energy transfer as in conventional heating.

Microwave heating has many advantages over conventional heating, including cost and energy savings, and considerable reduction in processing time [8]. By using microwave energy as a heating source, short sintering time at desired temperature offers an opportunity to control, especially the microstructure coarsening during sintering, leading to excellent mechanical properties [9]. Instead of using only microwaves as a heating source, microwave heating system through a combination of conventional conduction heating and energy conversion heating using a microwave is found to be more advantageous for heating or sintering of materials [10–12]. The advantages of microwave rapid sintering include rapid and more uniform heating, prevention of hot spot formation, and more uniform and finer microstructure leading to high-performance products [13, 14].

1.2. Microwave sintering of materials

In principle, sintering is one of the consolidation methods to make bulk objects from loose powder compacts by heating the material below its melting point. Conventionally, the green body (unsintered powder compact) is sintered using resistant heating. Since the resistive heating is the application of thermal energy, sintering process depends on the diffusion of atoms

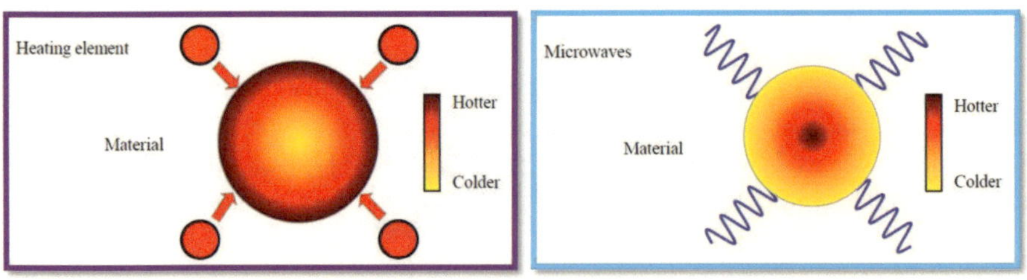

Figure 1. Heat distribution within a material during conventional and microwave heating.

that cause the adherence of loose particles to each other [13]. Sintering of materials using microwaves is a newly explored method, and it has been applied successfully in processing of various materials. For sintering using microwaves, the electromagnetic contrivance of the microwaves interact directly with the materials and, magnetic and dielectric losses lead to self-heating of materials. Initially, microwave energy was used to sinter various types of ceramic materials [9, 15]. By using microwave sintering, equal or superior performance ceramic products [16–18] can be produced through shorter sintering time at a lower temperature and at low cost when compared to the conventional sintered products. At room temperature, most of the ceramics do not couple well with microwaves particularly at 2.45 GHz microwave frequency and they are not heated appreciably [19]. Their coupling efficiency can be increased by increasing the temperature, using microwave susceptor/absorber, varying their morphology, and changing the microwave frequency. In most of the hybrid sintering methods, SiC of various forms such as SiC rods [4] with SiC sample holders [20] are preferably used mainly due to its high loss factor.

Most of the investigations [21–25] were conducted on the microwave sintering of semiconductors, inorganic, ceramics, and polymeric materials until 2000. Lack of research on microwave sintering of metals is based on the well-known fact that all metals reflect the microwaves causing arching during microwave heating and thus limited diffusion of the microwave waves. Later, researchers grabbed that arching phenomenon applies only for metal-based composites in the form of powder compacts [14, 26]. The idea of applying microwave energy to sinter metals and metallic materials is relatively new and limited studies on sintering of metal-based materials are available in the literature. From the literature, most of the studies are reported on the microwave sintering of iron-based materials and only a few reports are available on aluminum-, magnesium-, and copper-based materials/composites [14, 26–29].

2. Metal matrix composites

Development of metal matrix composites (MMCs) has been an important innovation in materials engineering over the past three decades. Metal matrix composites offer several attractive advantages over traditional engineering materials due to their superior properties [30, 31]. Metal matrix composites can be divided into three broad categories (**Figure 2**): (i) continuous fiber-reinforced matrix composites, (ii) small fiber-reinforced matrix composites, and (iii) particulate-rein-

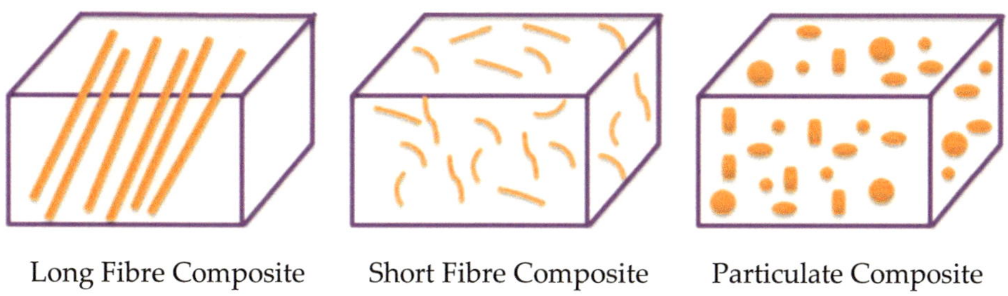

Figure 2. Types of reinforcement in a composite.

forced matrix composites. Among all these, particulate-reinforced metal matrix composites have gained decent interest because of their superior properties and low manufacturing expenditure. In light metal matrix composites (MMCs) [32], Al or Mg is mostly used as a base metal matrix, and ceramic particles (carbides, nitrides, and oxides) are generally used as reinforcement phase.

Nowadays, the demand of reducing energy consumption, especially in automotive industries, is becoming a critical issue. Development of lightweight aluminum-based composites can be considered as one of the promising solutions to address this issue.

The nano-sized reinforcements have a major role in improving the physical and mechanical properties which can be achieved by the addition of small volume fractions ($\leq$2%), whereas for micron-sized particle-reinforced metal matrix composites higher volume fractions ($\gg$10%) are essential [27]. Further addition of reinforcement will cause degradation of composite properties, which can be attributed to the possible agglomeration, clustering of reinforcement, and micro-porosity in the nanocomposites. Recently, there has been considerable interest in the production of metal matrix nanocomposites in which nanoparticulates are incorporated into the base matrix [33]. The production of nanocomposites is currently under exploration and is still at its laboratory scale research level. However, interestingly, when compared to composites with micron-sized reinforcements, nanocomposites exhibit comparable or better mechanical properties with the use of lesser amount of reinforcements [9–12]. Both casting and powder metallurgy (PM) methods can be used to fabricate metal matrix nanocomposites. Historically, PM methods have been developed successfully and commercially used by different manufactures and have also been applied in the production of MMCs for aerospace applications. As compared to casting methods, PM approach has shown its advantage to produce uniform microstructures leading to develop high-performance composite materials [34].

At present, the development of metal matrix composites with light metal matrices are gaining increasing attention due to their enhanced properties coupled with weight savings. These unique properties make them attractive for automotive and aircraft industries in which the weight reduction is the critical factor. So far, extensive studies have been done for the production of aluminum matrix composites and now these are being manufactured commercially for numerous industrial applications. The development of economical aluminum nanocomposites using cost-effective fabrication techniques will serve the requirement of the development of light-weight structural materials well suited to industrial and commercial applications. The main objective of our study was to fabricate high-performance aluminum metal matrix composites through cost-effective processing technique based on PM route incorporating microwave sintering method. A comparison of the microstructural, mechanical, and thermal properties is presented to elucidate the usefulness of the manufactured composite materials.

3. Fabrication of Al metal matrix composites

In the current book chapter, the presented composite materials were synthesized through the powder metallurgy method of mixing the matrix (pure aluminum) and reinforcements (SiC, Si_3N_4, and Al_2O_3). **Figure 3** presents the schematic flow chart of the experimental design. To produce Al-SiC nanocomposites, nano-sized SiC powder (1.5 vol.%) was added to pure Al.

http://dx.doi.org/ 10.5772/68081

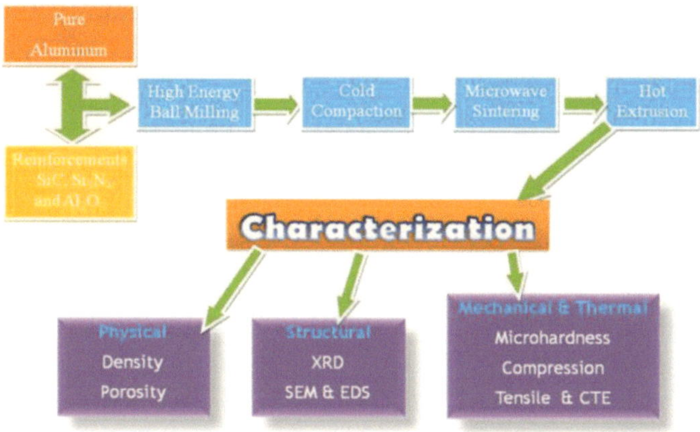

Figure 3. Schematic flow chart of the experimental design.

The blending of the mixture was carried out at room temperature using a Retsch PM400 planetary ball mill for 2 h with the milling speed of 200 rpm in order to get a homogeneous particle distribution. No balls were used in this stage. The ball milled powders were cold compacted using a uniaxial pressure of 50 tons into billets (40 mm length with 35 mm diameter). The sintering of the compacted cylindrical billets was carried out at 550°C using an innovative microwave sintering process [35], just below the melting temperature of Al. The other metal matrix composites (Al-1.5 vol.% Si_3N_4 and Al-15 vol.% Al_2O_3) were also prepared in a similar manner.

Prior to hot extrusion, the microwave sintered billets were soaked at 400°C for 1 h and then hot extruded at 350°C and 500 MPa. The extrusion ratio was ~20.25:1 to produce an 8 mm diameter extruded rod, as can be seen in **Figure 4(a)**. After extrusion, these rods were subsequently used for characterization studies.

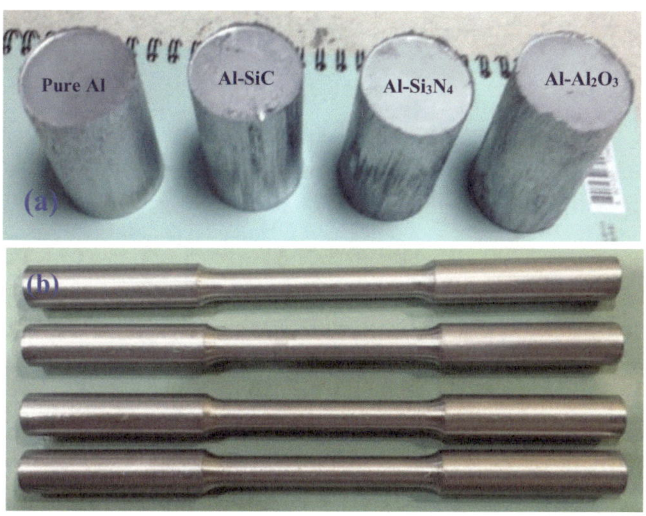

Figure 4. The pictures of the produced AMMCs.

The phase identification of the extruded samples was carried out using X-ray powder diffractometer (PANalytical X'pert Pro) based on Cu-K_α radiation (1.541 Å) in the 2θ range of 30–80° at scan rate of 0.2°/min. Individual phases were identified by matching the typical X-ray diffraction (XRD) peaks against JCPDS data. Field emission scanning electron microscopy (SEM) (JEOL JSM-6010 and Hitachi FESEM-S4300) with energy dispersion spectroscopy (EDS) was used to identify the reinforcement phase and microstructure of the extruded composite samples.

The hardness testing of the pure Al and composite samples was carried out using Vicker's hardness tester with applied load of 100 gf for 15 s as per the ASTM standard E384-08. Compressive testing of the cylindrical specimens was performed at room temperature according to the procedures outlined in ASTM standard E9-89a using Universal testing machine-Lloyd. Tensile testing of the extruded pure Al and its composite samples was done using a universal testing machine-Lloyd according to the ASTM E8/E8M-15a standard at room temperature under the strain rate of $8.3 \times 10^{-4}\ s^{-1}$. For tensile tests, round test specimens of 25 mm gauge length and 5 mm gauge diameter (**Figure 4(b)**) were prepared and tested on a fully automated servo-hydraulic mechanical testing machine, MTS-810. For every composition, three samples were tested to check repeatable values. The fractured surfaces of the selected compression and tensile specimens were studied by scanning electron microscope (Hitachi FESEM-S4300). Nanoindentation investigation was done using a MFP-3D Nanoidenter (head connected to AFM equipment) system equipped with standard Berkovich diamond indenter tip. The testing was performed at room temperature and the values of hardness (H) and Young's modulus (E) were directly obtained. The presented nanoindentation results are the average of six indentations values. Coefficients of thermal expansion of pure Al and developed composites were determined in the temperature range of 50–350°C using a INSEIS TMA PT 1000LT thermo-mechanical analyzer. A heating rate of 5°C/min was employed at argon flow rate of 0.1 lpm.

4. Properties of Al metal matrix composites

4.1. X-ray diffraction analysis of AMMCs

The X-ray diffraction (XRD) patterns for the microwave sintered and hot extruded pure Al and Al metal matrix with reinforcement particles SiC, Si_3N_4, and Al_2O_3 are shown in **Figure 5**.

The diffraction peaks of Al, SiC, Si_3N_4, and Al_2O_3 phases can be observed. The sharp peaks representing the presence of Al phase in the XRD patterns. The presence of SiC, Si_3N_4, and Al_2O_3 particles is indicated by minor peaks. The peaks of Al are indexed as (2 2 0), (3 1 1), (1 1 1), (2 0 0), and (2 2 2), whereas SiC peaks are as (1 1 1), (2 0 0); Si_3N_4 peak as (1 2 0); and Al_2O_3 peaks are as (0 1 2), (1 0 4), (1 1 3), (0 2 4), (1 1 6), (2 1 4), (3 0 0). During the microwave sintering and hot subsequent extrusion process, it is noted that no solid-state reaction took place between the matrix and reinforcement to form any other undesired phases. The XRD results also approve the elemental mapping results, as will be shown later in **Figure 7**, which verifies the fabrication of phase pure different ceramic-reinforced Al-composites.

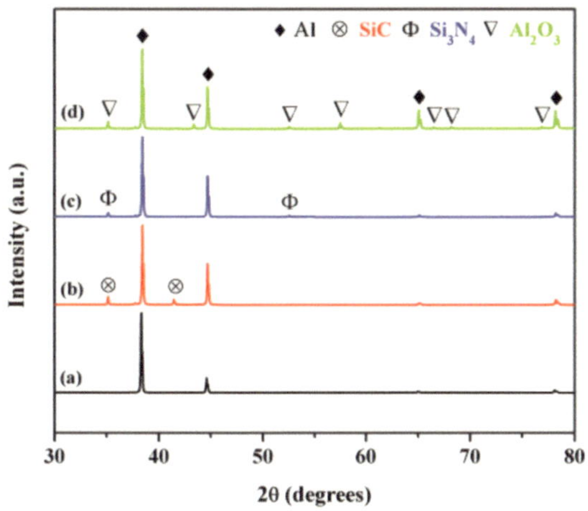

Figure 5. XRD patterns for (a) pure Al, (b) Al-SiC, (c) Al-Si_3N_4, and (d) Al-Al_2O_3 composites.

4.2. SEM analysis of AMMCs

Scanning electron microscopy (SEM) was used in order to analyze the morphology and microstructure of developed composites containing different reinforcements. **Figure 6** shows the SEM micrographs of microwave sintered-hot extruded pure Al and Al-X (X = SiC, Si_3N_4, and Al_2O_3) composites.

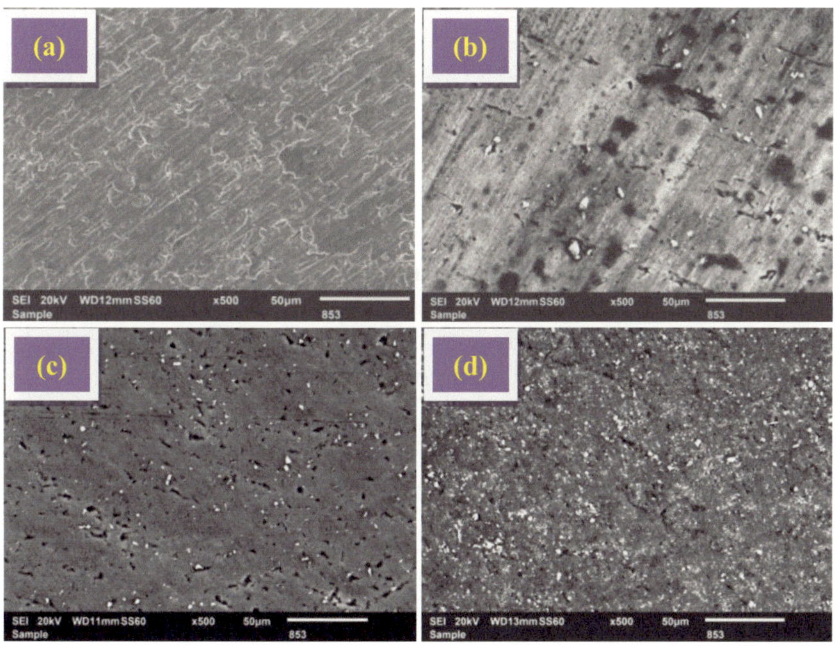

Figure 6. FESEM images for (a) pure Al, (b) Al-SiC, (c) Al-Si_3N_4, and (d) Al-Al_2O_3 composites.

The results revealed a fair uniform distribution of ceramic reinforcement particles in the aluminum matrix. It can be further noted that SEM images show two main phases: the grey matrix is the Al phase while the dispersed phase showing white spots represents the SiC, Si_3N_4, and Al_2O_3 particles used as reinforcements. At some spaces, agglomeration of reinforcement particulates has been observed (**Figure 6(d)**) which is due to density differences of reinforcements and the aluminum matrix.

The mechanical properties of Al-MMCs are dependent on the nature and distribution of the reinforcement particles. Homogeneous and intragranular distribution is preferred to attain improved properties, and importantly, the hot extrusion process has led to the desirable distribution. Previous studies have reported that agglomeration of reinforcement particles in Al matrix has resulted in the degradation of mechanical properties, as reinforcement clustering along with voids act as pre-existing cracks, limiting the stress transfer from soft matrix to hard phase particles during the deformation process [36, 37].

In our developed Al-composites, the agglomeration of reinforcements is observed only in a few locations, confirming a uniform reinforcement distribution in the Al-X composites. This near-uniform distribution of reinforcement promotes even heating (by absorbing the microwave energy) through the compact during sintering and demonstrates the effectiveness of using powder metallurgy and microwave sintering for the synthesis of Al-based composites [38]. Hence, the SEM results show that microwave sintering followed by hot extrusion process has an appropriate potential for manufacturing the ceramic particle-reinforced metal matrix composites.

4.3. EDS analysis of AMMCs

The energy dispersive spectroscopy (EDS) technique was used to study the composition and elemental distribution of phases present in the Al-based composites.

Figure 7(a–c) shows the EDS mapping analysis of microwave-hot extruded pure Al and Al-X (X = SiC, Si_3N_4, and Al_2O_3) composites. The elemental distribution of phases such as aluminum (matrix) and ceramic particles (reinforcements) is clearly observable. Furthermore, the reinforcing elements are uniformly dispersed all over the aluminum matrix. This confirms the appropriate mixing of ceramic reinforcement particles with the aluminum matrix. **Figure 7(a)** represents the corresponding Al, Si, and composition maps of the Al-SiC composite. **Figure 7(b)** shows the corresponding Al, Si, and N composition maps of the Al-Si_3N_4 composite. **Figure 7(c)** shows the EDS Al and O composition maps of the Al-Al_2O_3 composite. Further, the elemental distribution map evidently reveals the uniform distribution of ceramic reinforcement particles in Al matrix and confirms the presence of aluminum, SiC, Si_3N_4, and Al_2O_3 phases in respective composition.

4.4. Mechanical properties of AMMCs

The amount and size of reinforcement particles, type of processing technique, and the matrix/particle integrity greatly influence the mechanical properties of an Al-based composite. A strong matrix/particle interface integrity was obtained in this study. Therefore, the volume

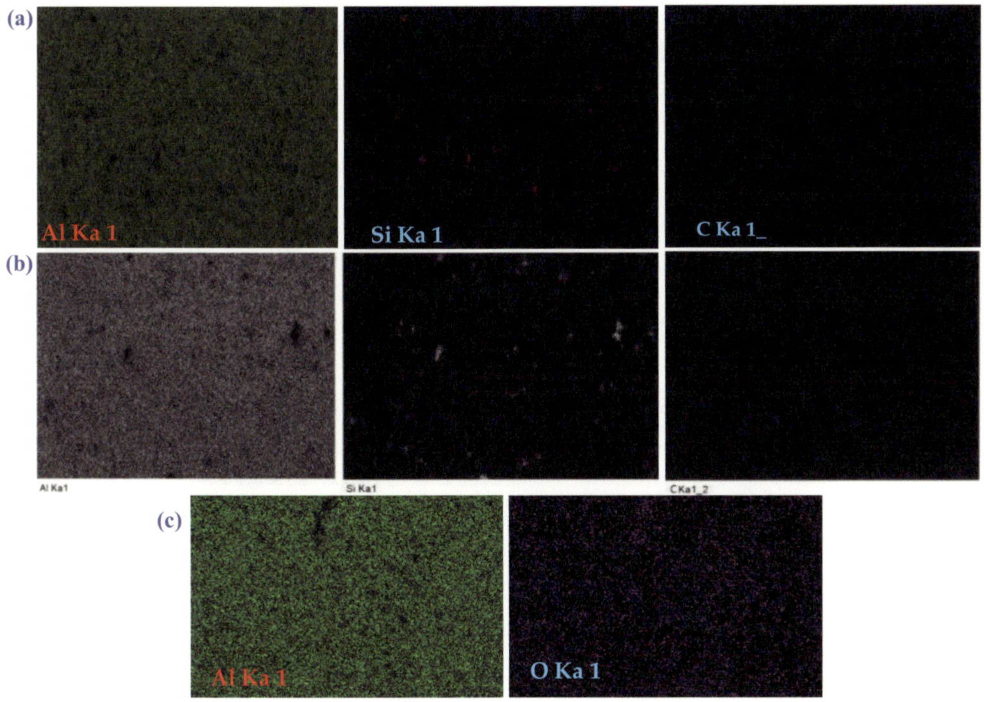

Figure 7. EDS maps of AMMCs reinforced with (a) SiC, (b) Si_3N_4, and (c) Al_2O_3.

fraction of the reinforced ceramic particles and the effect of hot extrusion, as a secondary processing, play an active role in improving the mechanical properties of the composites.

4.4.1. Microhardness studies of AMMCs

The microhardness is a very useful important property that reflects the strength of the material. Generally, several factors would affect the microhardness of the composites, such as particle shape, size, amount, distribution, density of reinforcement, and method of preparation [39].

Figure 8 shows the microhardness of the microwave sintered-extruded pure Al, SiC, Si_3N_4, and Al_2O_3 reinforced composite. The hardness of the Al-X (X = SiC, Si_3N_4, and Al_2O_3) composites is higher than the pure aluminum. The Al-1.5 vol.% SiC, Al-1.5 vol.% Si_3N_4, and Al-15 vol.% Al_2O_3 composites exhibit a hardness of 82 ± 4, 101 ± 3, and 92 ± 5 Hv, respectively; these values are comparatively higher than the unreinforced aluminum. However, a remarkable enhancement to 101 ± 3 is observed for Si_3N_4 reinforced Al composite. The increase in the microhardness of AMMCs indicates that the ceramic particles has a major contribution in the strengthening of Al matrix. This increase in the hardness is because of the contribution of the reduced crystallite size of ~15 nm (Si_3N_4) compared to 35 nm (SiC) in the composite.

The presence of hard ceramic particles can enhance the microhardness of composites according to the rule of mixtures [40].

$$H_c = H_m f_m + H_r f_r \tag{1}$$

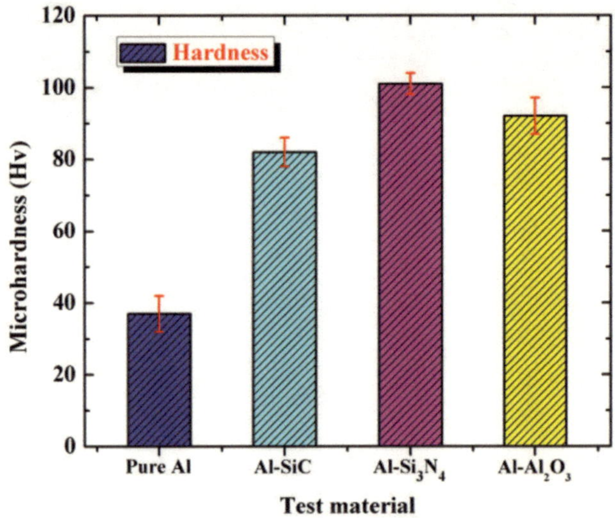

Figure 8. Hardness of aluminum metal matrix composites.

where H_c represents hardness of the composite, H_m and H_r represent hardness of the matrix and the reinforcing particle, respectively, and f_m and f_r represent the volume fraction of the matrix and the reinforcing particle, respectively.

The dispersion of hard ceramic reinforcement in the soft aluminum matrix results in strengthening of the structure. Referring to Hall-Petch relationship, the mechanical properties of the metallic materials are affected by the grain size. The grain size of metal matrix composites is smaller than that of the aluminum matrix because of grain refinement of reinforced ceramic particles. The fine grains enhance the hardness of the resulting structure. In addition, the difference in thermal shrinkage between the aluminum matrix and the ceramic particles produces quench hardening effect [41]. The presence of hard ceramic particles also improves the mechanical properties due to dispersion hardening of soft aluminum matrix. In fact, the presence of hard particles impedes the motion dislocation and thus improves the mechanical properties [42].

4.4.2. Compressive studies of AMMCs

The true stress-strain curves of the microwave sintered-hot extruded pure Al and Al-X (X = SiC, Si_3N_4, and Al_2O_3) composites under compression loading at room temperature are shown in **Figure 9**. The average compressive yield strength (CYS) and ultimate compressive strength (UCS) values of the extruded composites are listed in **Table 1**.

A significant improvement in the strength of Al-X composites are observed compared to pure aluminum. The compression strength of pure Al was increased by adding various ceramic reinforcement particles. The Al-1.5 vol.% SiC composite showed the compressive yield strength (0.2% CYS) and the ultimate compressive strength (UCS) of ~114 and ~392 MPa, respectively, the incremental increase is ~26 and ~72%, respectively, compared to pure Al. In the case of Al-1.5 vol.% Si_3N_4 composite, the (CYS) (UCS) were ~142 and ~412 MPa, respectively, showing

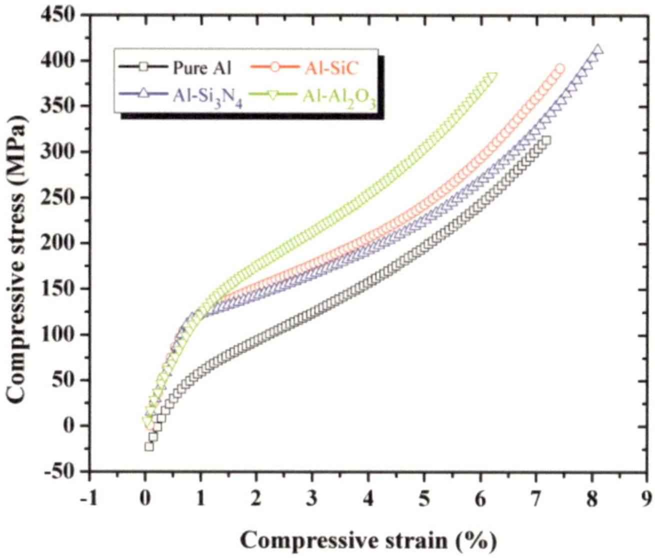

Figure 9. Compressive stress-strain curves of aluminum metal matrix composites.

an increase of ~31 and ~115%, respectively, compared to pure Al. The addition of micron-sized alumina particles, Al-15 vol.% Al_2O_3 composite exhibited (UCS) ~136 MPa and (CYS) 338 MPa which is ~24 and ~106%, respectively, higher than that of pure Al. The Al-X (X = SiC, Si_3N_4, and Al_2O_3) composites exhibited higher compressive failure strain values when compared to that of pure Al (~7.1%).

This significant improvement in compression strength properties of the extruded Al-X (X = SiC, Si_3N_4, and Al_2O_3) composites compared to the pure Al can be ascribed to the coupled effects of (i) uniform distribution of reinforcing particles in the matrix and (ii) enhanced dislocation density [39]. For a clearer comparison, we have noticed that the compressive properties of the microwave sintered-hot extruded Al-X (X = SiC, Si_3N_4, and Al_2O_3) composites are interestingly superior to that of conventional sintered AMMCs [43–47].

Materials	Nanoindentation data		Compressive properties			Tensile properties		
	Hardness (GPa)	Young's modulus (GPa)	CYS (MPa)	UCS (MPa)	Failure Strain (%)	TYS (MPa)	UTS (MPa)	Elongation (%)
Pure Al	5.15 ± 0.3	73 ± 5	70 ± 3	313 ± 5	7.17	105 ± 2	119 ± 4	13.6 ± 0.3
Al-1.5 vol.% SiC	9.60 ± 0.6	81 ± 6	114 ± 7	392 ± 6	7.48	158 ± 9	178 ± 6	7.3 ± 0.9
Al-1.5 vol.% Si_3N_4	16.34±0.4	94±2	142±6	412±3	8.07	165±5	191±5	8.2±0.4
Al-15 vol.% Al_2O_3	24.56 ± 0.8	106 ± 9	136 ± 5	388 ± 8	6.19	139 ± 8	154 ± 6	7.2 ± 0.7

Table 1. Mechanical properties of aluminum metal matrix composites.

4.4.3. Tensile studies of AMMCs

The representative tensile stress-strain curves for microwave sintered-hot extruded pure Al and Al-X (X = SiC, Si_3N_4, and Al_2O_3) composites at room temperature are shown in **Figure 10**. The variations in the tensile strength, yield strength, and ductility with reinforcement addition are listed in **Table 1**. It can be observed that all composites exhibited higher tensile strengths in comparison to that of pure Al. However, the elongation of the composites decreases as compared to pure Al. The calculated decrease is in elongation compared to microwave sintered-extruded pure Al is 46, 39, and 47% for Al-1.5 vol.% SiC, Al-1.5 vol.% Si_3N_4, and Al-15 vol.% Al_2O_3 composites, respectively.

Moreover, **Table 1** also shows that the tensile properties of the Al-1. 5 vol.% Si_3N_4 composites are comparable/superior to that of the SiC and Al_2O_3 reinforced Al composites. This can be endorsed to the reduced size of the reinforcing particles employed [37]. Like compressive properties, the tensile properties of microwave sintered-extruded Al-X (X = SiC, Si_3N_4, and Al_2O_3) composites are found superior to the conventional sintered AMMCs [31–35].

To understand the strengthening effects of ceramic reinforcement particles on the hardness, compression, and tensile properties of composites, such as UTS and YS, it is favorable to discuss the strengthening mechanism in detail. In the present study, strengthening occurs due to following mechanisms: (i) active load transfer from the matrix to the reinforcement, (ii) Orowan strengthening, and (iii) generation of internal thermal stresses because of the difference in the coefficient of thermal expansion (CTE) between the reinforcement particles and matrix phase.

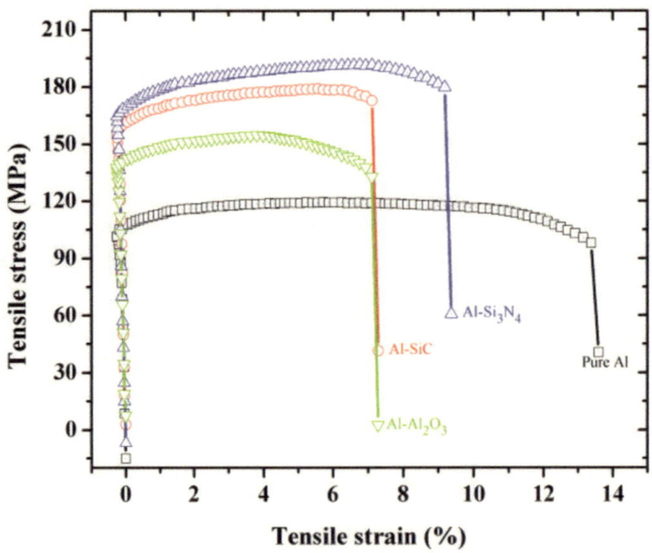

Figure 10. Tensile stress-strain curves of aluminum metal matrix composites.

The efficient load transfer (σ_{load}) between the ductile matrix and the hard ceramic reinforcement particles during tensile testing occurs, particularly when there is a good interfacial contact between the matrix and the reinforcement and is represented as following [48–50]:

$$\sigma_{load} = 0.5\, V_f \sigma_{YM} \tag{2}$$

where V_f is the volume fraction of ceramic reinforcement particles and σ_{YM} is the matrix yield stress.

The interaction between the dislocations and the reinforcement particles enhances the strength of the composite materials in agreement with the Orowan mechanism. Due to the existence of dispersed reinforcement particles in the matrix, dislocation loops are formed when dislocations interact with the reinforcing particles. σ_{Orowan} can be calculated as [51]:

$$\sigma_{Orowan} = \frac{0.13Gb}{\lambda} \ln \frac{r}{b} \tag{3}$$

where G is the shear modulus of matrix, b is the Burgers vector, λ is the inter-particle spacing, and r is the particle radius.

The difference in the CTE values of the reinforcement particles and the metal matrix produces geometrically necessary dislocations and thermally induced residual stresses. The thermal stresses at the particles and matrix interface make the plastic deformation more tough which, hence, enhances the level of hardness and flow stress. The mismatch strain effect due to the difference between the CTE values of particles and that of the matrix is given by [38]:

$$\Delta \sigma_{CTE} = \sqrt{3}\, \beta G_m b \sqrt{\frac{24\, V_f \Delta\alpha \Delta T}{(1 - V_f)\, b\, r_p}} \tag{4}$$

where b is the strengthening coefficient, $\Delta\alpha$ is the difference between CTE of matrix and reinforcement, and ΔT is the difference between the test and process temperature.

4.4.4. Fractography of AMMCs

Some selected compression and tensile tested fractured surfaces were studied using SEM in order to understand the type of fracture and nature of the bonding between the reinforcing particles and Al of the microwave sintered-extruded Al-X (X = SiC, Si_3N_4, Al_2O_3) composites.

The fracture morphology of microwave sintered-extruded pure Al and Al-X (X = SiC, Si_3N_4, Al_2O_3) composites during compression test are shown in **Figure 11(a–d)**. The fracture surfaces are comparatively smooth and the formation of shear band can barely be seen in the fractured samples. The fractured compressive samples reveal a crack at 45° to the test axis. **Figure 11(a–d)** shows a typical shear mode fracture in pure Al and Al-X (composites reinforced with various ceramic particles. It approves that the compressive deformation of the Al composites is expressively indifferent. This is due to heterogeneous deformation and work hardening behavior [52]. In contrast, mixed fracture surface and the shear band formation is found in Al-Al_2O_3

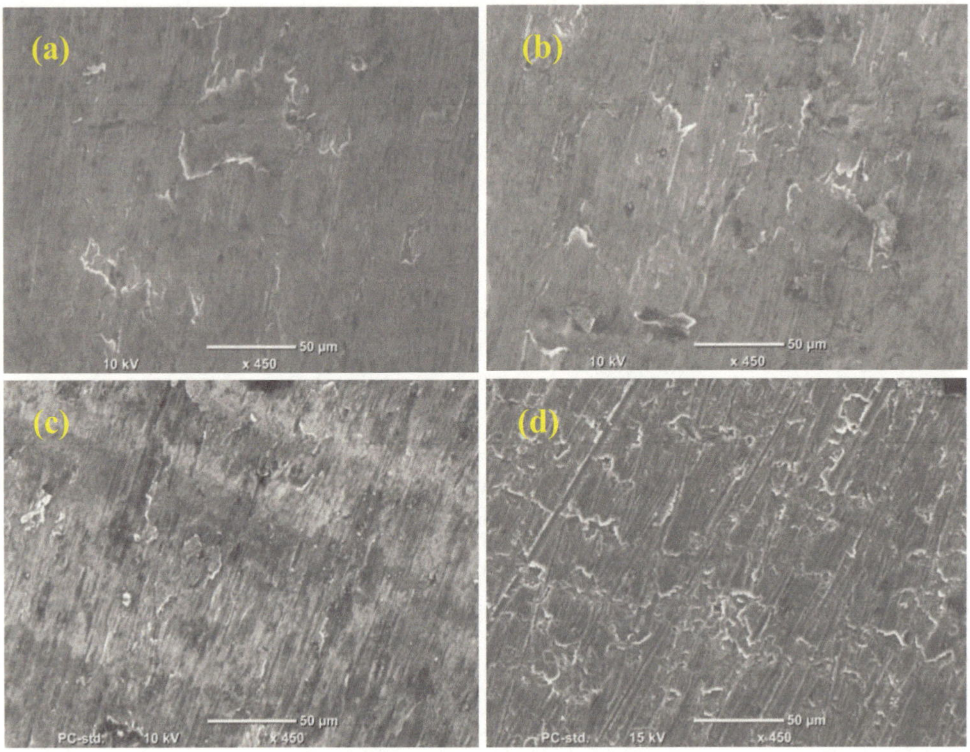

Figure 11. SEM micrographs of the compression fracture surfaces of (a) pure aluminum, (b)Al-SiC, (c) Al-Si_3N_4, and (d) Al-Al_2O_3.

composite (**Figure 11(d)**). The plastic deformation in the composites was inhibited due to the dispersion of second phases. This led to the significant reduction in compressive failure strain in the composite (see **Table 1**).

Fracture morphology of pure Al and its composites during tensile testing are presented in **Figure 12**. The examination of fractured surfaces reveals the formation of similar ductile fracture in all composites. For Al-SiC composites, the fractured surfaces shows dimple-like fracture which can be related to the observed failure strain of more than 7% (see **Table 1**). The presence of SiC particles in the dimple cores and walls suggests that the fractured particles and agglomerates are potential stress concentration sites and susceptible to void formation. Al-Si_3N_4 composites showed the strongest bonding as revealed by the good matrix/reinforcement performance. For Al-Al_2O_3 composites, it can be seen that ductile failure occurs in the matrix, whereas brittle, cleavage-type failure is seen to be predominant in regions where Al_2O_3 particles are present. Large number of dimples with tear ridges is also seen in the Al-Al_2O_3 composite.

4.4.5. Coefficient of thermal expansion of AMMCs

The variation of CTE of microwave sintered-extruded pure Al and Al-X composites (X = SiC, Si_3N_4, Al_2O_3) is shown in **Figure 13**. It can be observed that the CTE values decrease with reinforced ceramic particles. It is in accordance with the theory that the thermal expansion

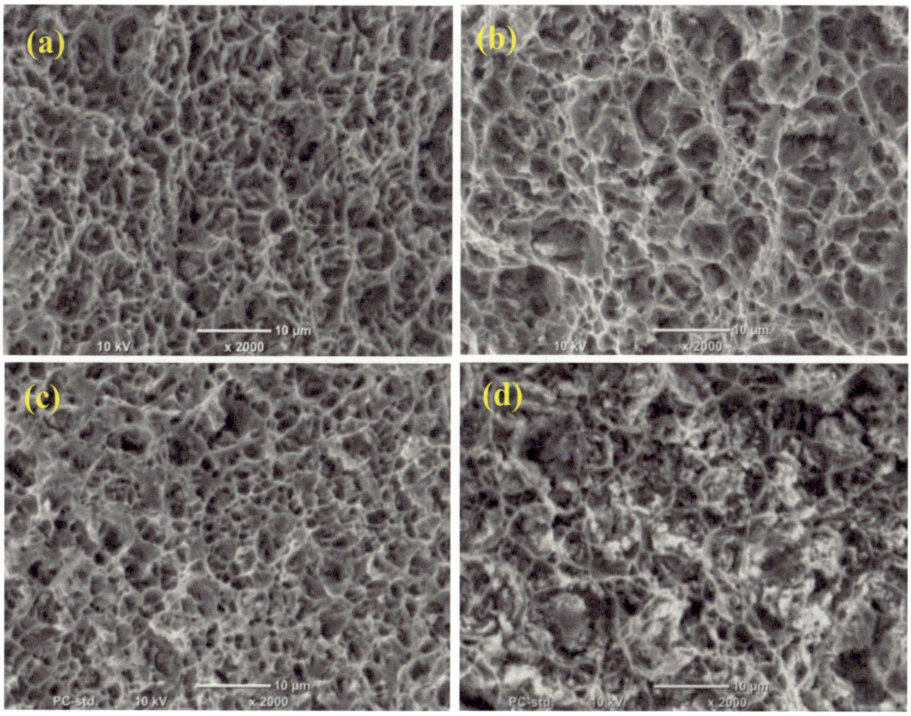

Figure 12. SEM micrographs of the tensile fracture surfaces of (a) pure aluminum, (b)Al-SiC, (c) Al-Si_3N_4, and (d) Al-Al_2O_3.

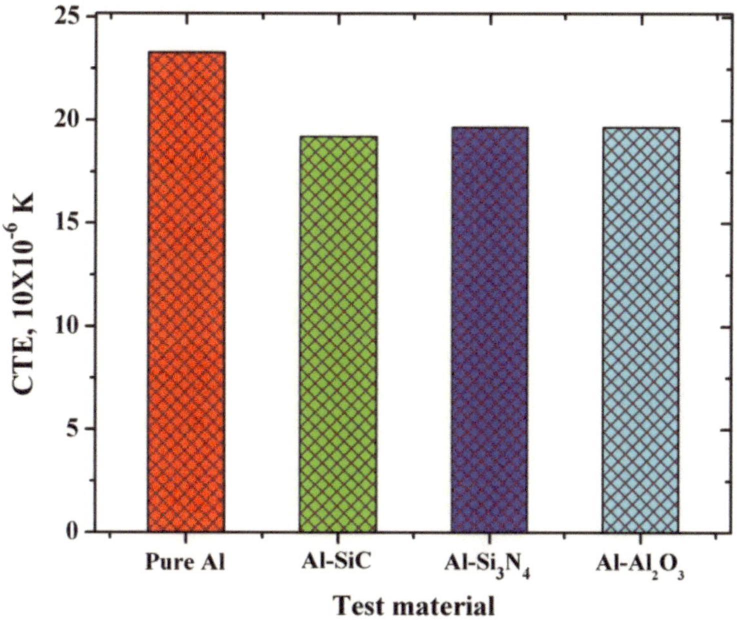

Figure 13. CTE of the aluminum metal matrix composites.

of the composites is governed by the competing interactions of expansion of Al matrix and the constraint of ceramic particles through their interfaces [52]. The CTE of pure Al was measured to be 23.31 × 10^{-6}/K which is in close agreement with the theoretical CTE of aluminum (24 × 10^{-6}/K). The addition of nano-sized 1.5 vol.% SiC, nano-sized 1.5 vol.% Si_3N_4, and micron-sized 15 vol.% Al_2O_3 particles to Al reduced the CTE value to ~19.20 × 10^{-6}/K, 19.43 × 10^{-6}/K, and 19.66 × 10^{-6}/K which are ~17.63, ~16.66, and ~15.65% reduction when compared to pure Al. This considerable decrease in CTE values may be due to the high thermal stability of SiC, Si_3N_4, and Al_2O_3 reinforcement particles having theoretical CTE of 4.3 × 10^{-6}/K [53], 3.3 × 10^{-6}/K [54], 7.4 × 10^{-6}/K [53], respectively. The linear decrease in CTE values with the addition of ceramic particles can be attributed to: (i) the lower CTE values of ceramic (SiC, Si_3N_4, and Al_2O_3) particles reinforcements as compared to that of the pure Al matrix; and (ii) uniform distribution of the ceramic reinforcements in the matrix.

The compatible CTE of Al-X composites (X = SiC, Si_3N_4, Al_2O_3) and high dimensional stability makes these microwave sintered-extruded composite very competitive for application in aerospace and automotive industry.

5. Conclusions

Al-X (X = SiC, Si_3N_4, and Al_2O_3) composites were successfully synthesized through microwave-assisted powder metallurgy route coupled with hot extrusion process. Various ceramic reinforcement particles were added into Al matrix, and their effect on structural, mechanical, and thermal properties has led to the following conclusions.

- XRD patterns of Al-X (X = SiC, Si_3N_4, Al_2O_3) composites indicate that the main components of the synthesized composites are Al, SiC, Si_3N_4, and Al_2O_3.
- Homogeneous reinforcement particles distribution was found in microwave sintered-extruded Al-X (X = SiC, Si_3N_4, Al_2O_3) composites. This shows that the microwave sintering coupled with extrusion process has an appropriate potential to process high-performance particle-reinforced metal matrix composites.
- A comparison of mechanical properties (hardness, strength) indicates that microwave sintered-extruded Al-X (X = SiC, Si_3N_4, Al_2O_3) composites have superior properties compared to microwave sintered-extruded pure Al. The improvement in mechanical properties can be attributed to (i) active load transfer from the matrix to the reinforcement, (ii) Orowan strengthening, and (iii) generation of internal thermal stresses because of the difference in the coefficient of thermal expansion (CTE) between the reinforcement particles and matrix phase.
- The produced Al-X (X = SiC, Si_3N_4, Al_2O_3) composites have low ductility compared to pure Al due to low inherent ductility of ceramic particles used as reinforcement.
- The fractography results indicate that under compressive loading, the Al-X (X = SiC, Si_3N_4, Al_2O_3) composites show the presence of shear bands which confirms the brittle mode of fracture. However, under tensile loading, the dimple formation was noticed on the fractured surfaces endorsing ductile mode of fracture.

- Coefficient of thermal expansion values decreases with the addition of ceramic-reinforced particles into Al matrix confirming high-dimensional stability of microwave sintered-extruded Al-X (X = SiC, Si_3N_4, Al_2O_3) composites making them suitable for automotive and many other related applications.

Acknowledgements

This chapter was made possible by NPRP Grant 7-159-2-076 from Qatar National Research Fund (a member of the Qatar Foundation). Statements made herein are solely the responsibility of the authors.

Author details

Penchal Reddy Matli[1], Rana Abdul Shakoor[1]* and Adel Mohamed Amer Mohamed[2]

*Address all correspondence to: shakoor@qu.edu.qa

1 Center for Advanced Materials, Qatar University, Doha, Qatar

2 Department of Metallurgical and Materials Engineering, Suez University, Suez, Egypt

References

[1] Sutton WH. Microwave processing of ceramic materials. American Ceramic Society Bulletin. 1989;**68**:376-386

[2] Sutton WH. Ceramic transactions microwaves: Theory and applications in materials processing II. Key issues in microwave process technology. 1993;**36**:3-18

[3] Das S, Mukhopadhyay AK, Datta S, Basu D. Prospects of microwave processing: An overview. Bulletin of Materials Science. 2009;**32**:1-13

[4] Xie Z, Wang C, Fan X, Huang Y. Microwave processing and properties of Ce-Y-ZrO_2 ceramics with 2.45 GHz irradiation. Materials Letters. 1999;**38**:190-196

[5] Agrawal DK. Microwave processing of ceramics: A review. Current Opinion in Solid State Materials Science. 1998;**3**:480-486

[6] Penchal Reddy M, Madhuri W, Ramamanohar Reddy N, Siva Kumar KV, Murthy VRK, Ramakrishna Reddy R. Magnetic properties of Ni-Zn ferrites prepared by microwave sintering method. Journal of Electroceramics. 2012;**28**:1-9

[7] Hao HS, Xu LH, Huang Y, Zhang XM, Xie ZP. Kinetics mechanism of microwave sintering in ceramic materials. Science in China Series E: Technological Sciences. 2009;**52**:2727-2731

[8] Oghbaei M, Mirzaee O. Microwave versus conventional sintering: A review of fundamentals, advantages and applications. Journal of Alloys and Compounds. 2010;**494**:175-189

[9] Roy R, Agrawal D, Cheng J, Gedevanishvili S. Full sintering of powdered-metal bodies in a microwave field. Nature. 1999;**399**:668-670

[10] Lianxi H, Erde W. Fabrication and mechanical properties of SiCw/ZK51A magnesium matrix composite by two-step squeeze casting. Materials Science and Engineering A. 2000;**278**:267-271

[11] Jia DC. Influence of SiC particulate size on the microstructural evolution and mechanical properties of Al-6Ti-6Nb matrix composites. Materials Science and Engineering A. 2000;**289**:83-90

[12] Tang F, Hagiwara M, Schoenung JM. Formation of coarse-grained inter-particle regions during hot isostatic pressing of nanocrystalline powder. Scripta Materialia. 2005;**53**: 619-624

[13] Jenkins I, Wood J. Powder Metallurgy: An Overview. London: Institute of Metals; 1991

[14] Gupta M, Wong WLE. Microwaves and Metals. Singapore: John Wiley & Sons Pte. Ltd.; 2007

[15] Penchal Reddy M, Ubaid F, Shakoor A, Mohamed AMA, Gupta M. Microwave rapid sintering of Al-metal matrix composites: A review on the effect of reinforcements, microstructure and mechanical properties. Metals. 2016;**6**:1-19

[16] Rajkumar K, Aravindan S. Microwave sintering of copper graphite composites. Journal of Materials Processing Technology. 2009;**209**:5601-5605

[17] Leonelli C, Veronesi P, Denti L, Gatto A, Iuliano L. Microwave assisted sintering of green metal parts. Journal of Materials Processing Technology. 2008;**205**:489-496

[18] Binner J, Annapoorani K, Paul A, Santacruz I, Vaidhyanathan B. Dense nanostructured zirconia by two stage conventional/hybrid microwave sintering. Journal of the European Ceramic Society. 2008;**28**:973-977

[19] Menezes RR, Souto PM, Kiminami RHGA. In: Lakshmanan A, editor. Sintering of Ceramics—New Emerging Techniques Microwave Fast Sintering of Ceramic Materials. Croatia: InTech; 2012. pp. 3-26

[20] Madhuri W, Penchal Reddy M, Ramamanohar Reddy N, Siva Kumar KV, Murthy VRK. Comparison of initial permeability of MgCuZn ferrites sintered by both conventional and microwave methods. Journal of Physics D: Applied Physics. 2009;**42**:165007

[21] Bykov Yu V, Rybakov KI, Semenov VE. High-temperature microwave processing of materials. Journal of Physics D: Applied Physics. 2001;**34**:R55-R75

[22] Agrawal D. Microwave sintering of ceramics, composites, metals, and transparent materials. Journal of Materials Education. 1999;**19**:49-58

[23] Agrawal DK, Papworth AJ, Cheng J, Jain H, Williams DB. Microstructural examination by TEM of WC/Co composites prepared by conventional and microwave processes. Powder metallurgical high performance materials. Proceedings. 2001;**2**:677-684

[24] Rodiger K, Dreyer K, Gerdes T, Porada MW. Microwave sintering of hardmetals. International Journal of Refractory Metals & Hard Materials. 1998;**16**:409-416

[25] Agrawal D, Cheng J, Seegopaul P, Gao L. Grain growth control in microwave sintering of ultrafine WC/Co composite powder compacts. Powder Metallurgy. 2000;**43**:15-16

[26] Penchal Reddy M, Ubaid F, Shakoor A, Mohamed AMA, Madhuri W. Structural and mechanical properties of microwave sintered Al-$Ni_{50}Ti_{50}$ composites. Journal of Science Advanced Materials and Devices. 2016;**1**:362-366

[27] Wong WLE, Gupta M. Simultaneously improving strength and ductility of magnesium using nano-size SiC particulates and microwaves. Advanced Engineering Materials. 2006;**8**:735-739

[28] Sankaranarayanan S, Shankar VH, Jayalakshmi S, Bau NQ, Gupta M. Development of high performance magnesium composites using $Ni_{50}Ti_{50}$ metallic glass reinforcement and microwave sintering approach. Journal of Alloys and Compounds. 2015;**627**:192-199

[29] Mula S, Panigrahi J, Kang PC, Koch CC. Effect of microwave sintering over vacuum and conventional sintering of Cu based nanocomposites. Journal of Alloys and Compounds. 2014;**588**:710-715

[30] Aigbodion VS, Hassan SB. Effects of silicon carbide reinforcement on microstructure and properties of cast Al-Si-Fe/SiC particulate composites. Materials Science and Engineering A. 2007;**447**:355-360

[31] Sharifi EM, Karimzadeh F, Enayati MH. Fabrication and evaluation of mechanical and tribological properties of boron carbide reinforced aluminum matrix nanocomposites. Materials and Design. 2011;**323**:263-3271

[32] Kainer KU. Basic of Metal Matrix Composites. In: Kainer KU, editor. Metal matrix composites: Custom-made materials for automotive and aerospace engineering. Weinheim, Chichester: Wiley-VCH; 2006

[33] Ye J, He J, Schoenung JM. Cryomilling for the fabrication of a particulate B_4C reinforced Al nanocomposite: Part I. Effects of process conditions on structure. Metallurgical and Materials Transactions A. 2007;**37**:3099-3109

[34] Tjong SC. Novel nanoparticle-reinforced metal matrix composites with enhanced mechanical properties. Advanced Engineering Materials. 2007;**9**:639-652

[35] Gururaj P, Vyasaraj M, Ganesh Kumar M, Gupta M. Enhancing the hardness/compression/damping response of magnesium by reinforcing with biocompatible silica nanoparticulates. International Journal of Materials Research. 2016;**107**:1091-1099

[36] Sun C, Song M, Wang Z, He Y. Effect of particle size on the microstructures and mechanical properties of SiC-reinforced pure aluminum composites. Journal of Materials Engineering and Performance. 2011;**20**:1606-1612

[37] Chawla N, Williams JJ, Saha R. Mechanical behavior and microstructure characterization of sinter-forged sic particle reinforced aluminum matrix composites. Journal of Light Metals. 2002;**2**(4):215-227

[38] Wong WLE, Gupta M. Using microwave energy to synthesize light weight/energy saving magnesium based materials: A review. Technologies. 2015;**3**:1-18

[39] Khadem SA, Nategh S, Yoozbashizadeh H. Structural and morphological evaluation of Al-5 vol% SiC nanocomposite powder produced by mechanical milling. Journal of Alloys and Compounds. 2011;**509**:2221-2226

[40] Roy R, Agrawal D, Cheng J, Mathis M. Microwave processing: Triumph of applications-driven science in WC-composites and ferroic titanates. Ceramics Transactions. 1997;**80**:3-26

[41] Dinaharan I, Sathiskumar R, Murugan N. Effect of ceramic particulate type on microstructure and properties of copper matrix composites synthesized by friction stir processing. Journal of Materials Research and Technology. 2016;**5**:302-316

[42] Sug WK, Lee UJ, Han SW, Kim DK, Ogi K. Heat treatment and wear characteristics of Al/SiCp composites fabricated by duplex process. Composites: Part B. 2003;**34**:737-745

[43] Ezatpour HR, P0061rizi MT, Sajjadi, SA. Microstructure and mechanical properties of extruded Al/Al_2O_3 composites fabricated by stir-casting process. Transactions of Nonferrous Metals Society of China. 2013;**23**:1262-1268

[44] Davis LC, Andres C, Allison JE. Microstructure and strengthening of metal matrix composites. Materials Science and Engineering A. 1998;**294**:40-45

[45] Sharma P, Sharma S, Khanduja D. Production and some properties of Si_3N_4 reinforced aluminum alloy composites. Journal of Asian Ceramic Societies. 2015;**3**:352-359

[46] Ma ZY, Li YL, Liang Y, Zheng F, Bi J, Tjong SC. Nanometric Si_3N_4 particulate-reinforced aluminum composite. Materials Science and Engineering A. 1996;**219**:229-231

[47] Zakeri M, Vakili-Ahrari Rudi A. Effect of shaping methods on the mechanical properties of Al-SiC composite. Materials Research. 2013;**16**(5):1169-1174

[48] Zhang Z, Chen DL. Contribution of Orowan strengthening effect in particulate reinforced metal matrix nanocomposites. Materials Science and Engineering A. 2008;**483**:148-152

[49] Zhang Z, Chen DL. Consideration of Orowan strengthening effect in particulate reinforced metal matrix nanocomposites: A model for predicting their yield strength. Scripta Materialia. 2006;**54**:1321-1326

[50] Habibnejad K, Mahmudi R, Poole WJ. Enhanced properties of Mg-based nano composites reinforced with Al_2O_3 nano-particles. Materials Science and Engineering A. 2009;**519**:198-203

[51] Ashby MF. Proceedings of the second bolton landing conference on oxide dispersion strengthening. New York: Gordon and Breach, Science Publishers Inc.; 1968

[52] Anggara B, Soegijono SB. Mechanical properties of metal Al/SiC and AlCu/SiC metal matrix composites (MMCs). KnE Engineering. 2016;1-5

[53] Brandes EA, Brook GB, Smithells Metals Reference Book, Butterworth-Heinemann, London, 1998, pp. 5-12

[54] Qing CG, Shu YW, Kang MA, Hussain M, Tao JL, Hui WUG. Aging and thermal expansion behavior of Si_3N_4p/2024Al composite fabricated by pressure infiltration method. Transactions of Nonferrous Metals Society of China. 2011;**211**:262-273

Chapter 3

Sintering of the Tricalcium Phosphate-Titania-Magnesium Fluoride Composites

Ibticem Ayadi and Foued Ben Ayed

Additional information is available at the end of the chapter

http://dx.doi.org/10.5772/intechopen.68501

Abstract

Titania (TiO_2) and magnesium fluoride (MgF_2) can be mixed with tricalcium phosphate (β-$Ca_3(PO_4)_2$, β-TCP) to make bioceramic composites, which would combine the biocompatibility of the β-TCP and the high tribological properties of TiO_2 and MgF_2 for biomedical applications. The samples were characterized by different characterization techniques such as physicochemical and mechanical. The sintering of the TCP at various temperatures (1000, 1100, 1200 and 1300°C) with different percentages of titania (2.5, 5, 7.5, 10, 20, 30, 40 and 50 wt%) was studied. The performances of the TCP-TiO_2 composites increase with both the sintering temperature and the amount of titania. The highest values of the composites' (H = 270 Hv; E = 33.1 GPa and G = 15.7 GPa) were obtained with 40 wt% titania at 1200°C. Moreover, the addition of 4 wt% MgF_2 to the TCP-40 wt% TiO_2 composites leads to better mechanical properties (σ_r = 27 MPa; H = 360 Hv; E = 51 GPa and G = 20 GPa) at 1200°C for 1 h. The amelioration of these properties is due to the formation of a new compound and the liquid phase which helps to fill the pores in the microstructure. The obtained performances of the TCP-TiO_2-MgF_2 composites are similar to those of bone tissue and especially as enamel.

Keywords: sintering, biomaterial, composites, mechanical properties, tricalcium phosphate

1. Introduction

Tricalcium phosphate (β-$Ca_3(PO_4)_2$, β-TCP) based on biomaterial has attracted considerable interest for orthopedic and dental applications [1–13]. The tricalcium phosphate has been used clinically to repair bone defects for many years [6–10]. However, the major limitation of the β-TCP uses as load bearing biomaterial is their poor fatigue resistance [6, 9]. Moreover, the mechanical properties of tricalcium phosphate are generally inadequate for many load-carrying applications [3, 6, 9]. Hence, inert oxides such as alumina (Al_2O_3) or zirconia (ZrO_2) have been widely studied due to their bio-inertness, excellent tribological properties, fracture toughness and strength [6, 9, 11–15]. Thus, the study conducted by Sakka et al. has recently concerned with the tricalcium phosphate/alumina system [11]. The researchers were interested in producing the Al_2O_3-TCP composites with different percentages of β-TCP (10, 20, 40 and 50 wt%). In fact, the best mechanical properties of this study reached 13.5 MPa with a composition of the alumina-10 wt% tricalcium phosphate composite after the sintering at 1600°C [11]. Thereby, the next study directed by Sakka et al. proved that the incorporation of 5 wt% TiO_2 to the alumina-10 wt% tricalcium phosphate composite matrix leads to the highest mechanical performances (74 MPa) at 1600°C [12]. However, the very low content of tricalcium phosphate (10 wt%) in the composites based on alumina, titania and tricalcium phosphate limits its use in the biomedical applications. Furthermore, the presence of the sizeable grains hinders the performances of these composites caused by the high temperature sintering [12]. Also, the study made by Sallemi et al. was interested to elaborate and to characterize the tricalcium phosphate-zirconia composites with different percentages of zirconia (25, 50 and 75 wt%) [14]. Thus, the ultimate values of the performances of the tricalcium phosphate were obtained with 50 wt% zirconia [14]. But, the experimental results in this study indicate the appearance of low mechanical properties of the tricalcium phosphate-zirconia composites [13, 14]. Based on the results, the inverse allotropic transformation of zirconia is sufficient to cause the degradation of mechanical properties of these composites [13, 14]. Thus, the expansion in volume of the zirconia samples is responsible for the fragility of the tricalcium phosphate-zirconia composites [13, 14]. In conclusion, the addition of alumina or zirconia to the TCP matrix did not enhance the mechanical properties [11–14]. For this reason, we tested another inert oxide like titania (TiO_2). Besides, titania has attracted much attention due to its excellent ability to chemically bond with living hard tissue [16, 17]. This happens through the formation of a bone mineral, such as tricalcium phosphate phase on the material surface that ultimately induces direct bonding with native bone tissue [18–20]. Thus, titania has been attracting attention as an implantable material because it is harmless to a living body, and with good mechanical properties. Moreover, titania is an attractive material, applicable to various fields, such as biomedical applications [16–20]. Furthermore, the titania is considered as having excellent biocompatibility, manifested in various biomedical applications [16–20]. So, the choice of the titania added to the tricalcium phosphate matrix was based on those considerations. Thereby, we would combine the biocompatibility of the TCP with the high mechanical and tribological properties of titania in order to elaborate a bioceramic composite. This combination of calcium phosphate/inert oxide system could give rise to more biomaterials in the physiological environment [21]. Few papers were interested on the study of the TCP-TiO_2 composites using hydroxyapatite and titania as starting materials [22–24]. In fact, titania has been added to the tricalcium phosphate matrix in the order of enhancing the mechanical performances of the TCP and not degrading its biocompatibility. Thereby, this study focused on the sintering and the mechanical properties of the tricalcium phosphate with different percentages of TiO_2 (2.5, 5, 7.5, 10, 20, 30, 40 and 50 wt%). In this study, the performances of the TCP-TiO_2 composites increase with the augmentation of the sintering temperature and the amount of titania. Thus, the best mechanical properties resulting from this study were obtained with the TCP-40 wt% TiO_2 composites after the sintering process at 1200°C for 1 h. The optimum values of the

TCP-40 wt% TiO_2 composites of mechanical strength, Vickers hardness, Young's and Shear modulus reached 33 MPa, 270 Hv, 33.1 GPa and 15.7 GPa, respectively. Thus, the aim of this study is to ameliorate the performances of the TCP-40 wt% TiO_2 composites by the inclusion of several additives to reach mechanical property similar to those of the bone tissues and precisely the enamel. Hence, MgF_2 has been chosen as a suitable compound for doping calcium phosphate [25]. Magnesium fluoride is an important material, which attracts attention thanks to its wide range of applications [26–32]. In the light of its technological properties, magnesium fluoride was studied in different area [26, 30, 33]. Among the alkaline-earth fluorides, MgF_2 crystallizes in the rutile-type structure [26, 27]. Thanks to the higher compressibility of MgF_2 and the relative sizes of its ions, the phase transformations of this material are found at lower pressures than those in many oxides [32]. The minor ions (Mg^{2+} and F^-) are incorporated in either the hydroxyapatite (HAp) or TCP crystal structure to replace the position of calcium, phosphorus or hydroxyl ion [25]. Fluoride is known to be important in suppressing dental caries [33, 34]. It stimulates the proliferation and the differentiation of bone cells [35]. The fluorine ion (F^-) has been investigated as an essential element for bone and dental formation in the human body [35, 36]. Fluoride compound is known as an effective additive for reducing the phase decomposition of hydroxyapatite due to the crystal structure stabilization [37]. Magnesium is an abundant and essential cation in the human body, since it has significant effects on human metabolism [38]. Mg^{2+} promotes the dental caries formation caused by the high calcification process in the bone formation [37, 38]. Even though, the lack of Mg^{2+} ion results from the inhibition of bone growth, the degradation of bone structure and the decrease of osteoblast adhesion [37, 38]. Evis and Pinar Sun [34] demonstrate the β-TCP structure stabilization and the hydroxyapatite formation by the heat treatment above 800°C. Hot pressed magnesium fluoride has been used as a dome material for many devices due to its excellent mechanical strength and thermal stability [39–41]. The study conducted by Kim et al. aimed to increase the mechanical properties and to inhibit the phase decomposition of HAp by the addition of MgF_2 in the HAp/ZrO_2 composites [37]. As a result, the MgF_2 (5 and 10 vol%) added to the HAp-zirconia composites completely suppresses the decomposition of hydroxyapatite into TCP below 1400°C owing to the ion substitution F^- for OH^- in the hydroxyapatite (HAp) crystal structure [37]. Many studies show that, the addition of MgF_2 to the tetragonal zirconia and HAp mixtures strongly reduces the tendency of HAp decomposition due to partial substitution of F^- for OH^- ions in the calcium phosphate structure [42, 43]. Therefore, MgF_2 enhances the tricalcium phosphate-titania composites performances by the inclusion of fluoride (F^-) and magnesium (Mg^{2+}) ions. On the other hand, fluoride presents the much more stable ion in acidic environment (pH = 4). This property is very important in medical fields and especially in dentistry [44]. According to the previous studies, fluoride has a great influence on the physical and biological properties of materials [45–47]. Additionally, fluoride is an essential trace element required for normal dental and skeletal development [47]. It has been shown that the presence of fluoride offers beneficial effects on increasing the quantity and quality of bone formation in the body [47, 48]. In the second time, we are interested in the examination of the effect of magnesium fluoride addition (1, 2.5, 3, 4, 4.5, 5, 6, 7.5 and 10 wt%) on the performances of the TCP-40 wt% TiO_2 composites sintered at 1200°C for different lengths of the sintering time. Then, we will characterize the resulting composites by different techniques such as Brazilian test (σ_r), Vickers indentation and ultrasonic technique, the magic angle scanning nuclear magnetic resonance (^{31}P), the scanning electron microscopy, the infrared spectroscopy and the X-ray diffraction (XRD).

2. Materials and methods

The β-TCP powder resulted from a mixture of calcium carbonate ($CaCO_3$: Fluka, purity ≥ 98.5%) and calcium phosphate dibasic anhydrous ($CaHPO_4$: Fluka, purity ≥ 99%) after a heat treatment at 1000°C for 2 h according to the following reaction (Eq. (1))" [49]:

$$2CaHPO_4(s) + CaCO_3(s) \leftrightarrow Ca_3(PO_4) + H_2O + CO_2(g) \quad (1)$$

The initial powders used to obtain the TCP-TiO_2-MgF_2 composites were synthesized tricalcium phosphate (β-TCP), magnesium fluoride MgF_2 (Sigma Aldrich, purity > 98%) and titania TiO_2 (Fluka, purity > 98%). Firstly, Titania was introduced in the β-TCP matrix at different contents (2.5, 5, 7.5, 10, 20, 30, 40 and 50 wt%) followed by homogeneous mixing in a mortar. The heat treatment of the compacted disks was carried out in a vertical programmable muffle furnace (Pyrox 2408) at various temperatures (1000, 1100, 1200 and 1300°C) for 1 h. The theoretical density (d) was determined using the following Eq. (2):

$$d = (3.07A + 3.89B)/100 \quad (2)$$

where A and B are the weight ratios and 3.07 and 3.89 are the theoretical densities of β-TCP and anatase-TiO_2, respectively. The calculated theoretical densities of each composite are illustrated in **Table 1**.

Different amounts of MgF_2 (1, 2.5, 3, 4, 4.5, 5, 6, 7.5 and 10 wt%) were added to the TCP-40 wt% TiO_2 composites followed by homogeneous mixing in a mortar. The heat treatment of the compacted specimens was carried out in a vertical programmable muffle furnace (Pyrox 2408) at 1200°C for different sintering times. The theoretical density (d) was determined using the following Eq. (3):

A[a] (wt%)	B[a] (wt%)	d[b]
2.50	97.50	3.0905
5.00	95.00	3.1110
7.50	92.50	3.1315
10.00	90.00	3.1520
20.00	80.00	3.2340
30.00	70.00	3.3160
40.00	60.00	3.3980
50.00	50.00	3.4800

[a]A and B are the weight rates of TiO_2 and β-TCP, respectively.
[b]Theoretical density.

Table 1. The weight ratios of different TCP-TiO_2 composites.

$$d = (3.07\,A' + 3.89\,B' + 3.15\,C')/100 \quad (3)$$

where A′, B′ and C′ are the weight ratios and 3.07, 3.89 and 3.15 are the theoretical densities of β-TCP, anatase-TiO_2 and MgF_2, respectively. The calculated theoretical densities of all-composite are illustrated in **Table 2**.

The powders were mixed with absolute ethanol in an agate mortar. After milling these powders, the mixture was dried at 80°C for 24 h. After drying, the powder mixtures were molded in a cylinder with diameter 20 mm and thickness 4 mm and pressed under 150 MPa. The specimens were heated and cooled at rates of 10°C min^{-1} and 20°C min^{-1}, respectively.

The determination of the phase transformation in the microstructure of the elaborated composites was investigated by X-ray diffraction (XRD) analysis. The identification of the components phases was done by means Seifert XRD 3000 TT diffractometer with CuK_α radiation (λ = 1.54056 Å). The phase identification was operated resulting from the comparison between experimental XRD patterns and standards files compiled by the International Center for Diffraction Data (ICDD). The powders were then characterized by infrared spectrometric analysis with attenuated total reflection method (ATR) (Agilent Cary 630 Fourier Transform Infrared Spectrometer (FTIR)). The powders were submitted to the ^{31}P magic angle scanning nuclear magnetic resonance (MAS-NMR) on a Brucker 300 WB spectrometer.

The microstructure of the fractured surfaces of the sintered samples was investigated with scanning electron microscope (SEM) (JEOL JSM 5800LV) after enhancing their conductivity with a gold layer.

The powder's size was measured through a Micromeritics Sedigraph 5000. The specific surface area (SSA) was determined by the BET method with azote (N_2) as the adsorbed gas (ASAP 2010) [50].

A′[a] (wt%)	B′[a] (wt%)	C′[a] (wt%)	d[b]
59.50	39.50	1.00	3.3947
58.75	38.75	2.50	3.3897
58.50	38.50	3.00	3.3881
58.00	38.00	4.00	3.3848
57.75	37.75	4.50	3.3831
57.50	37.50	5.00	3.3815
57.00	37.00	6.00	3.3782
56.25	36.25	7.50	3.3732
55.00	35.00	10.00	3.3650

[a]A′, B′ and C′ are the weight rates of β-TCP, TiO_2 and MgF_2, respectively.
bTheoretical density.

Table 2. The weight ratios of the different TCP-TiO_2-MgF_2 composites.

The main particle size (D_{BET}) was calculated by assuming that the primary particles are spherical (Eq. (4)) [51]:

$$D_{BET} = \frac{6}{S\rho} \tag{4}$$

where ρ is the theoretical density and S is the surface specific area.

The densification of the sintered specimens was evaluated by the measurements of the specimen dimensions. The relative error of the densification value was about 1%.

Thermal expansion-shrinkage of the compact powder was measured with a dilatometer (Model Setaram TMA 92 dilatometer). About 20 mg of powder was heated with a heating rate of 10°C min^{-1} along with alumina as a reference material and a low stream of argon gas. The heating and cooling rates were 10 and 20°C min^{-1}, respectively.

The Brazilian test was used to measure the mechanical strength of the samples [52, 53]. The rupture strength (σ_r) (or mechanical strength) was calculated based on the maximum applied load recorded (P) and sample's dimensions (D being the diameter and t being the thickness) using (Eq. (5)):

$$\sigma_r = \frac{2 \times P}{\pi \times D \times t} \tag{5}$$

The sintered samples were examined by Vickers test indentation using loading values of 49 or 98 N applied for 15 s. This test was investigated after polishment of the samples surfaces between 1 and 3 μm with diamond paste. Vickers hardness value (H) was determined by Eq. (6) [54]:

$$H = 1.854(P/d^2) \tag{6}$$

where H is Vickers hardness, 'd' is the indent diagonal length and P is the indentation load.

The characterization of the samples using ultrasound technique promotes the determination of both Young's and Shear moduli value [55]. The measurement was done by a high-frequency generator Model 5077PR (Olympus). Young's modulus and the Shear modulus were calculated from the point of the longitudinal and the transversal ultrasonic velocities [55].

3. Results

3.1. Characterization of the starting powders

Table 3 presents the particles size distribution data (measured by granulometric repartition), the SSA and the calculated average grain sizes (D_{BET}) of the β-TCP, TiO_2 and MgF_2 powders. The averages of the grain's size of the samples obtained through the SSA (D_{BET}) and the granulometric repartition (D_{50}) are different. This difference may be due to the presence of agglomerates in the initial powders (**Table 3**).

The curve of the differential thermal analysis of the β-TCP shows the presence of two endothermic peaks: the first large band located between 1230 and 1270°C is attributed to the first allotropic transformation of the tricalcium phosphate (β–α) and the second peak registered at 1470°C is relative to the second transformation of TCP (α–α') (**Figure 1a**). These results are mentioned in previous works from literature [12, 56]. The Differential thermal analysis (DTA) curve of the TiO_2 shows that no evolution can be obtained with the sintering temperature (**Figure 1b**). The thermogram of the TCP-10 wt% TiO_2 composites reveals the appearance of two endothermic peaks (**Figure 1c**). The first peak at 1470°C is attributed to the second transformation of the β-TCP (α–α') while the new peak at 1440°C is probably relative to the

Compounds	SSA (m^2/g)	D_{BET} (μm)	D_{50}[a] (μm)	d[b]
TiO_2	12.00	0.11	0.20	3.89
β-TCP	0.80	2.80	6.00	3.07
MgF_2	37.00	0.05	0.10	3.15

[a]Mean diameter.
[b]Theoretical density.

Table 3. Characteristics of the different powders used in the study.

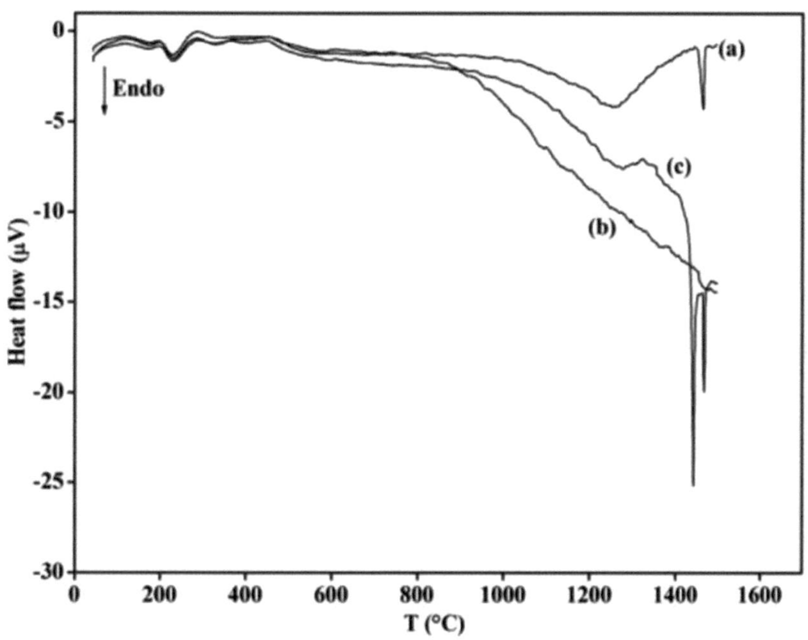

Figure 1. DTA curves of: (a) β-TCP, (b) anatase-TiO_2 and (c) TCP-10 wt% TiO_2 composites.

liquid phase formation. Furthermore, the melting points of both β-TCP and TiO_2 are 1756 and 1855°C, respectively [22]. Thus, the liquid phase is formed between TiO_2 and β-TCP, and it is not relative to the melting point of the initial powders. The results are similar to that found in the previous studies [12, 22]. They show that in the TiO_2/β-TCP system, there is an

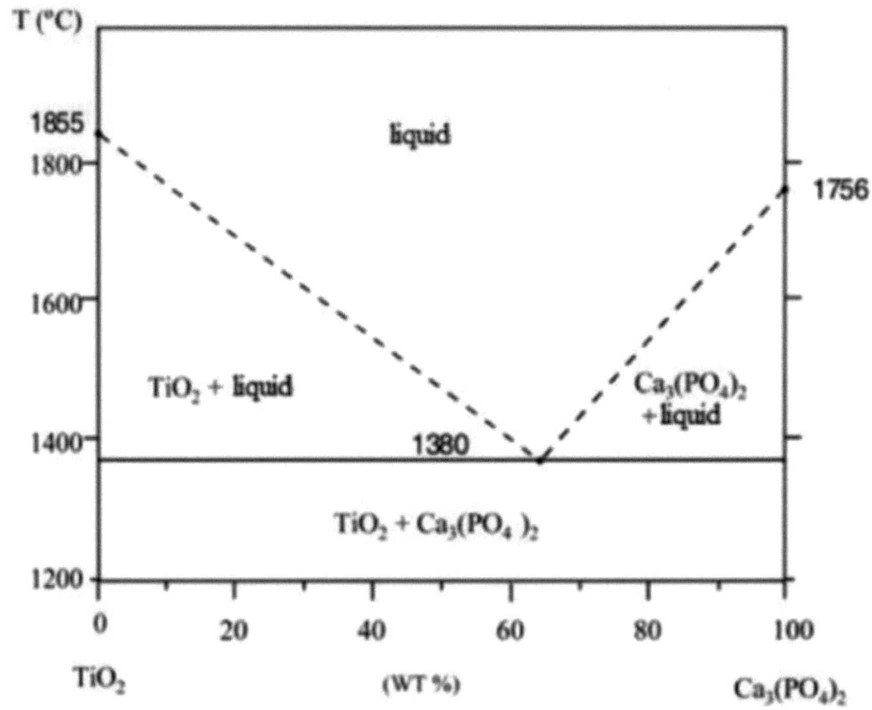

Figure 2. Phase equilibrium diagram of the TiO_2-$Ca_3(PO_4)_2$ system [22].

eutectic with a composition of 63 wt% TCP-37 wt% TiO_2 at 1380°C [22]. **Figure 2** confirms the presence of the liquid phase which is showed in the binary diagram of the $Ca_3(PO_4)_2$/TiO_2 [22].

Figure 3 shows the dilatometric measurements of different powders used in this study (β-TCP, TiO_2, TCP-10 wt% TiO_2 composites and MgF_2). The sintering temperature of the initial powder began at about 1000, 900 and 800°C for the β-TCP, the TiO_2 and the MgF_2, respectively (**Figure 3a–c**). The peak at 1230°C is attributed to the first allotropic transformation of the tricalcium phosphate (**Figure 3a**). The shrinkage curve of the titania powder reveals one peak which is relative to the phase transformation from anatase to rutile at 1090°C (**Figure 3b**). This result was well-confirmed by literature [57]. In fact, they showed that the commercial anatase-titania powder has been transformed into the rutile structure at about 1000°C [57], which confirms our results. The curve of the pure MgF_2 indicates that the shrinkage was manifested at 800°C until 1150°C (**Figure 3c**). In fact, the rate of maximum densification corresponds to the inflection point, which is obtained at 890°C (**Figure 3c**). **Figure 3d** presents the shrinkage curve of the TCP-10 wt% TiO_2 composites. In fact, no evolution was reported with the TCP-10 wt% TiO_2 composites (**Figure 3d**). As a result, the content of 10 wt% TiO_2 stabilizes the TCP structure and prevents the inverse allotropic transformation of TCP from the α phase to the β phase. The stabilization was well-explained by the exchange of Ca^{2+} and Ti^{4+} ions derived from β-TCP and TiO_2.

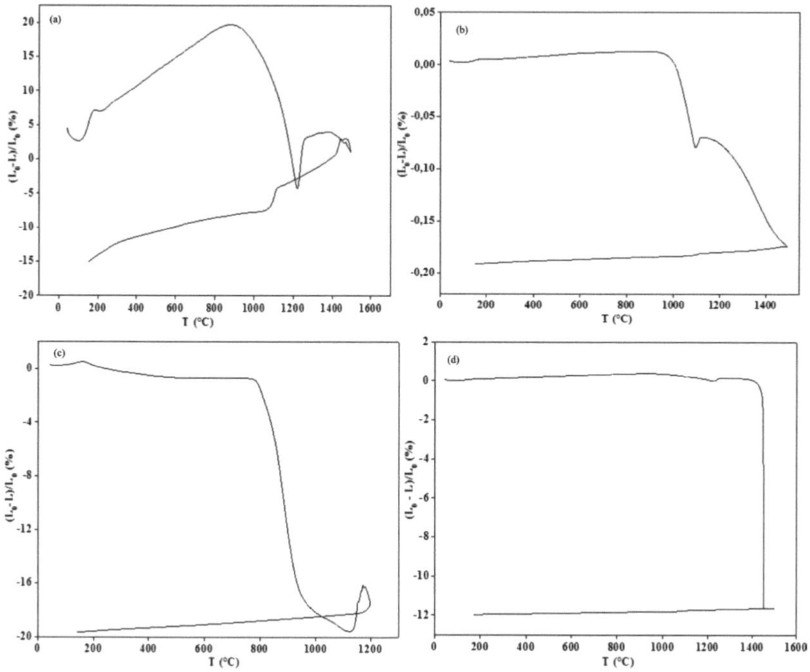

Figure 3. Linear shrinkage curves of: (a) β-TCP, (b) TiO_2, (c) MgF_2 and (d) TCP-10 wt% TiO_2 composites.

3.2. Effect of the addition of titania on the densification and the mechanical properties of the tricalcium phosphate

The evolution of the densification of the tricalcium phosphate was studied with the addition of titania between 1000 and 1300°C. **Figure 4** shows the typical relationship between temperature and density. The density of the β-TCP sintered with different percentages of the TiO_2 (2.5, 5, 7.5, 10, 20, 30, 40 and 50 wt%) increases with the sintering temperature (**Figure 4**). At 1200°C, the optimum value of the densification (89%) was obtained with 40 wt% TiO_2 (**Figure 4g**). Above 1200°C, the performances of the composites decrease abruptly (**Figure 4**).

Figure 5 shows the influence of the titania additive (2.5, 5, 7.5, 10, 20, 30, 40 and 50 wt%) at various sintering temperatures (1000, 1100, 1200 and 1300°C) on the rupture strength of TCP. The mechanical strength of the TCP-TiO_2 composites improves with both the content of TiO_2 and the sintering temperature (**Figure 5**). At 1200°C, the rupture strength of the TCP-40 wt% TiO_2 composites reached its maximum value (33 MPa) (**Figure 5g**). Above 1200°C, the rupture strength of the TCP-TiO_2 composites was hindered abruptly (**Figure 5**). The discrepancy of results appears especially after the addition of 50 wt% TiO_2 in the tricalcium phosphate matrix (**Figure 5h**). The amelioration of these performances of the TCP-40 wt% TiO_2 composites could be according to the formation of a liquid phase between TiO_2 and β-TCP.

The evolution of Vickers hardness with different percentages of titania at various temperatures was shown in **Figure 6**. Vickers hardness reached its optimum value (270 Hv) at 1200°C with 40 wt% TiO_2. Then, Vickers hardness value decreases with the sintering temperature.

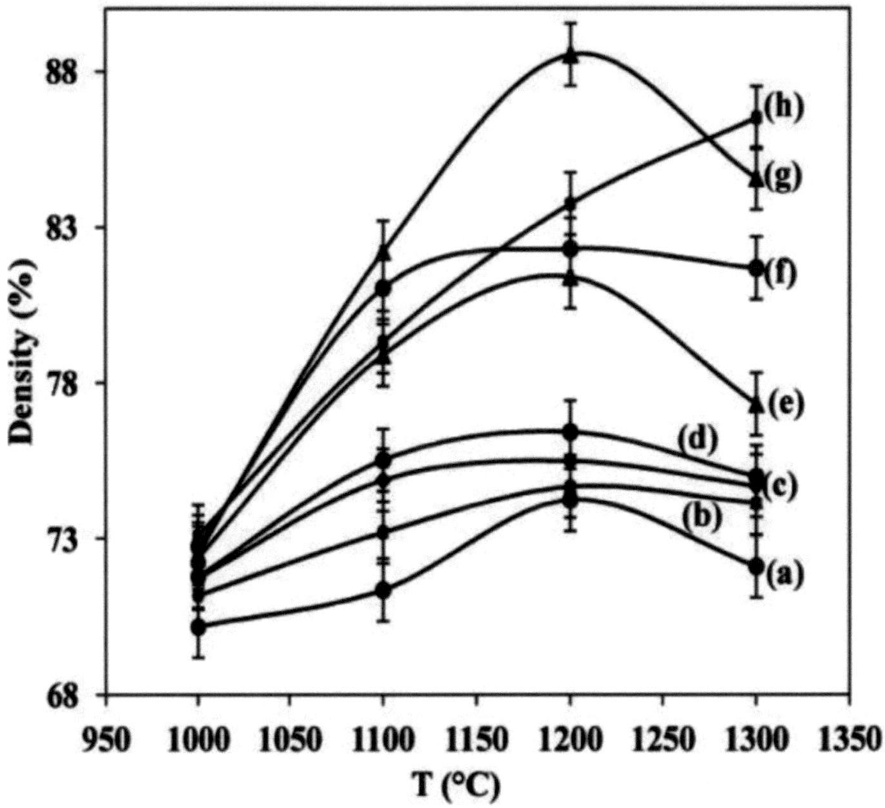

Figure 4. Relative density versus temperature of the β-TCP sintered for 1 h with different percentages of TiO_2: (a) 2.5 wt%, (b) 5 wt%, (c) 7.5 wt%, (d) 10 wt%, (e) 20 wt%, (f) 30 wt%, (g) 40 wt% and (h) 50 wt%.

Figures 7 and **8** present the evolution of elastics moduli (E and G) of the β-TCP sintered with different contents of TiO_2 (2.5, 5, 7.5, 10, 20, 30, 40 and 50 wt%) at various temperatures (1000, 1100, 1200 and 1300°C). The optimum values of both Young's modulus and Shear modulus reached 33.1 and 15.7 GPa, respectively, with 40 wt% TiO_2. Beyond 40 wt% TiO_2, the properties of the composites were hold up with the increase of the sintering temperature (**Figures 7** and **8**).

3.3. Effect of the magnesium fluoride addition on the densification and the mechanical properties of the tricalcium phosphate-titania composites

The effect of the magnesium fluoride addition on the performances of the TCP-40 wt% TiO_2 composites has been assessed by the measurement of the density and the rupture strength. The densification behavior of the TCP-40 wt% TiO_2 composites sintered at 1200°C for 1 h with different amounts of MgF_2 (1, 2.5, 3, 4, 4.5, 5, 6, 7.5 and 10 wt%) is reported in **Figure 9**. These results show a significant improvement of the relative density of the TCP-TiO_2 composites as a function of the magnesium fluoride added (**Figure 9**). The density of the TCP-TiO_2 composites increases from 3 wt% MgF_2 and remains constant until 6 wt%

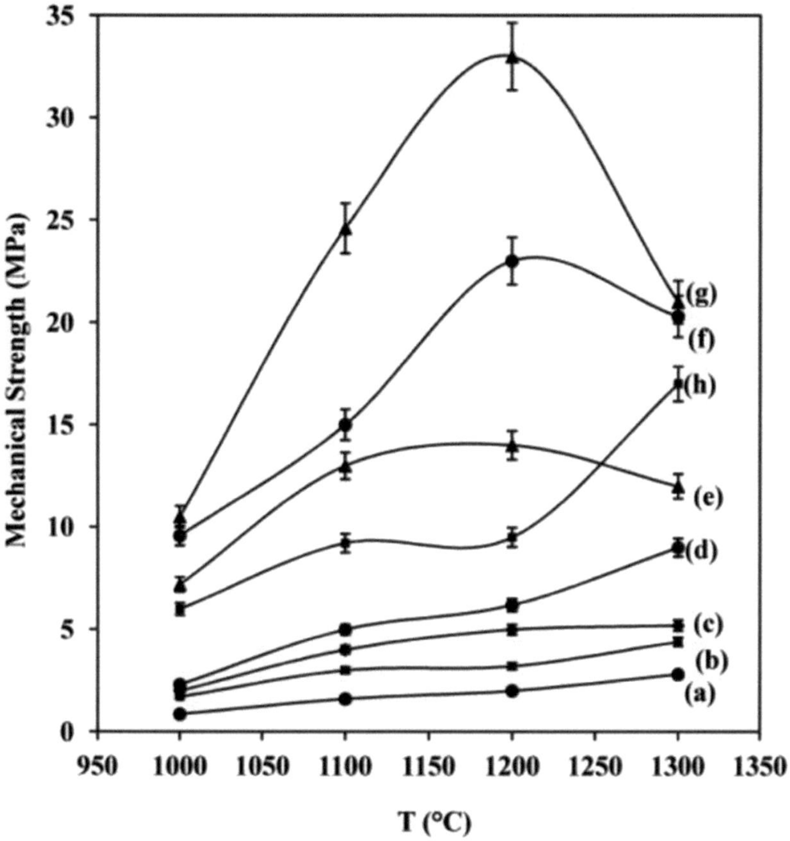

Figure 5. Mechanical strength versus temperature of the β-TCP sintered for 1 h with different percentages of TiO_2: (a) 2.5 wt%, (b) 5 wt%, (c) 7.5 wt%, (d) 10 wt%, (e) 20 wt%, (f) 30 wt%, (g) 40 wt% and (h) 50 wt%.

MgF_2, then the density of the composites decreases abruptly (**Figure 9**). The optimum value of the densification was about 94% with 4 wt% MgF_2. This result is obviously close to the full densification.

Figure 10 shows the rupture strength of the TCP-40 wt% TiO_2 composites sintered at 1200°C for 1 h with different percentages of MgF_2 (1, 2.5, 3, 4, 4.5, 5, 6, 7.5 and 10 wt%). Thus, the rupture strength reached its maximum value (27 MPa) after the addition of 4 wt% MgF_2 (**Figure 10**). Above 4 wt% MgF_2, the rupture strength decreases sharply (**Figure 10**). A remarkable amelioration of the rupture strength of the TCP-40 wt% TiO_2 composites was obtained with 10 wt% MgF_2 (**Figure 10**).

The influence of the magnesium fluoride addition on Vickers hardness was studied at 1200°C with the same percentages of MgF_2 (**Figure 11**). The addition of MgF_2 outstandingly enhanced Vickers hardness of the TCP-40 wt% TiO_2 composites (**Figure 11**). Thus, the incorporation of 4 wt% of MgF_2 to the composites led to a maximum of Vickers hardness (360 Hv) after the

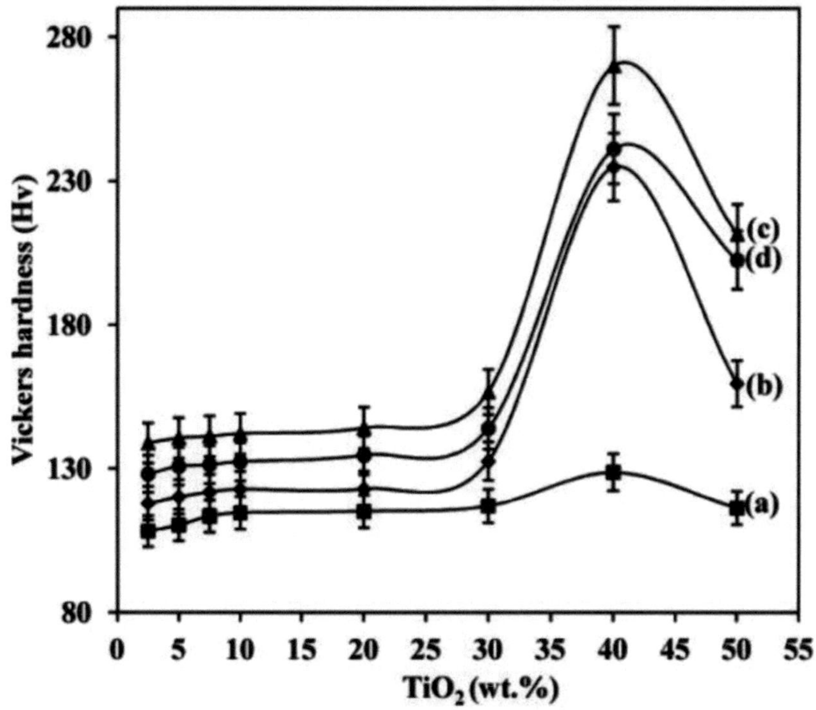

Figure 6. Vickers hardness versus temperature of the β-TCP sintered for 1 h with different percentages of TiO_2 at: (a) 1000°C; (b) 1100°C; (c) 1200°C and (d) 1300°C.

sintering process at 1200°C for 1 h. Beyond 4 wt% MgF_2, Vickers hardness of the composites was hindered (**Figure 11**). However, Vickers hardness of the TCP-40 wt% TiO_2 composites increases slowly with 10 wt% MgF_2 (**Figure 11**).

The evolution of the elastic modulus (E and G) of the TCP-40 wt% TiO_2 composites sintered with different contents of MgF_2 is shown in **Figure 12A** and **B**, respectively. The elastic modulus of the TCP-40 wt% TiO_2 composites increase with the addition of MgF_2 (**Figure 12A** and **B**). Thus, the optimum values of Young's modulus and the Shear modulus of the TCP-40 wt% TiO_2 composites are obtained by adding 4 wt% MgF_2 and these values reach 51 and 20 GPa, respectively. The performances of the samples were hindered with the increase of the percentages of MgF_2 in the TCP-40 wt% TiO_2 composites (**Figure 12A** and **B**).

Figure 13 illustrates a typical relation between the different lengths of the sintering time and the densification of the TCP-38 wt% TiO_2-4 wt% MgF_2 composites after the sintering temperature at 1200°C. The maximum value of densification (94%) was registered after the sintering process at 1200°C for 1 h (**Figure 13**). The densification's value decreases with the sintering time (**Figure 13**).

Figure 14 depicts the rupture strength evolution of the samples after the sintering process at 1200°C for different length of the sintering time. The mechanical strength of the TCP-TiO_2-MgF_2

http://dx.doi.org/10.5772/intechopen.68501

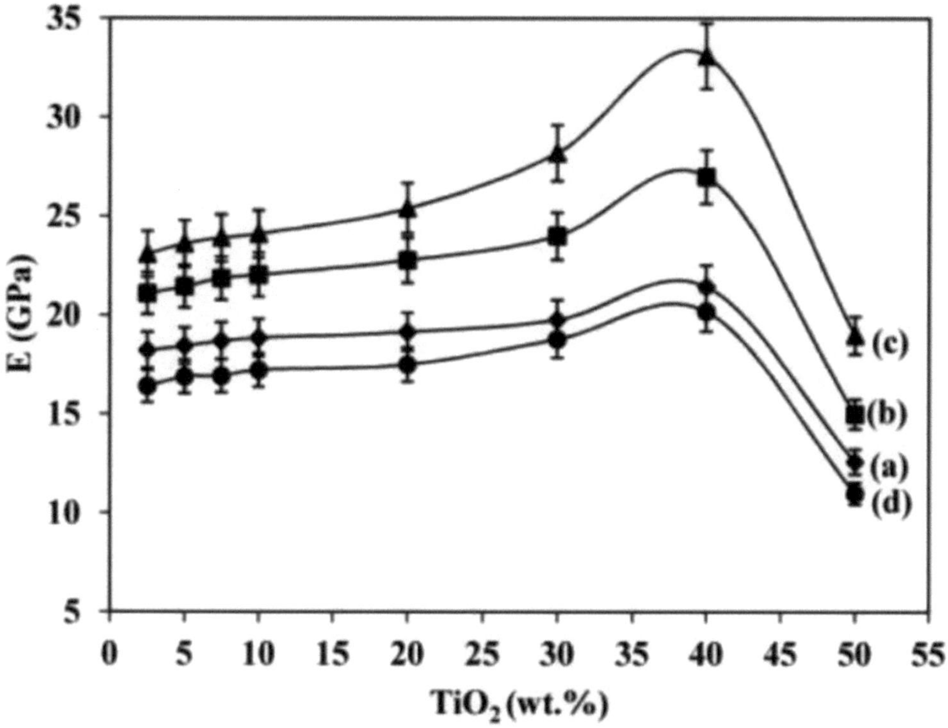

Figure 7. Young's modulus of the β-TCP sintered for 1 h with different percentages of TiO_2 at: (a) 1000°C; (b) 1100°C, (c) 1200°C and (d) 1300°C.

composites was enhanced after the sintering process at 1200°C for 1 h and reached 27 MPa then, decreases abruptly (**Figure 14**). This amelioration was associated to the important densification (**Figures 13** and **14**).

Figure 15 shows the evolution of Vickers hardness of the TCP-38 wt% TiO_2-4 wt% MgF_2 composites with the length of the sintering time. Vickers hardness increases and reaches its optimum value (360 Hv) for 1 h (**Figure 15**). Then, a remarkable diminution of Vickers hardness of the samples was registered above 1 h (**Figure 15**).

The evolution of the elastic moduli (E and G) of the TCP-38 wt% TiO_2-4 wt% MgF_2 composites sintered for different length of the sintering time at 1200°C is pictured in **Figure 16A** and **B**, respectively. The sintering time has no effect on the elastic moduli (**Figure 16A** and **B**). Young's modulus reached 51 GPa as maximum value while Shear modulus reached 20 GPa as an optimum value (**Figure 16A** and **B**).

3.4. Characterization of the TCP-TiO_2-MgF_2 composites

After the sintering process, the samples were investigated by different characterization techniques such as: XRD, IR, ^{31}P MAS-NMR and SEM.

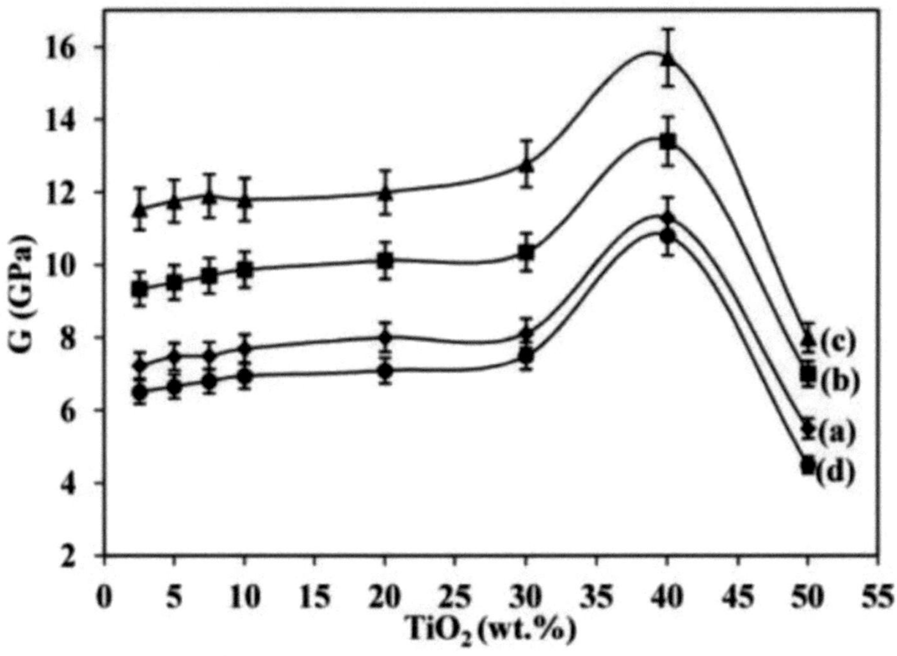

Figure 8. Shear modulus of the β-TCP sintered for 1 h with different percentages of TiO_2 at: (a) 1000°C, (b) 1100°C, (c) 1200°C and (d) 1300°C.

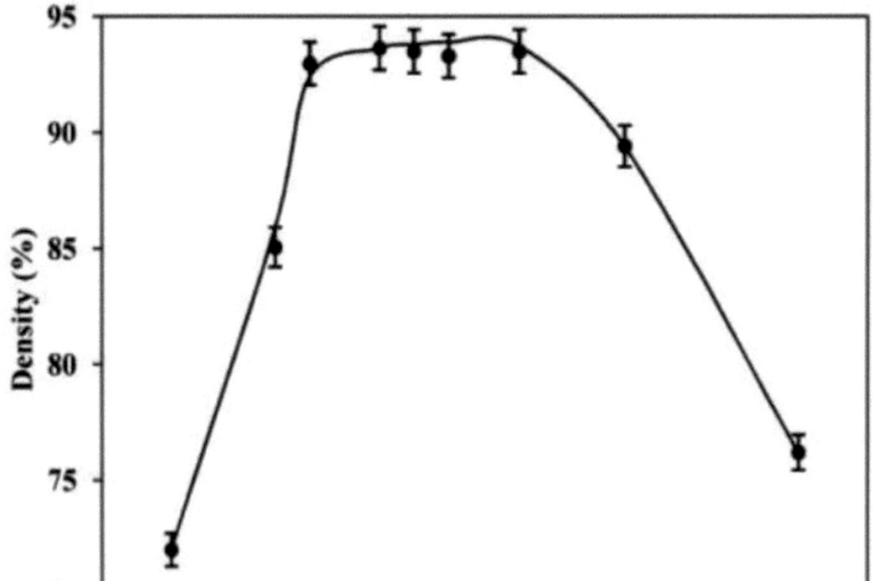

Figure 9. Relative density of the TCP-40 wt% TiO_2 composites sintered at 1200°C for 1 h with different percentages of MgF_2.

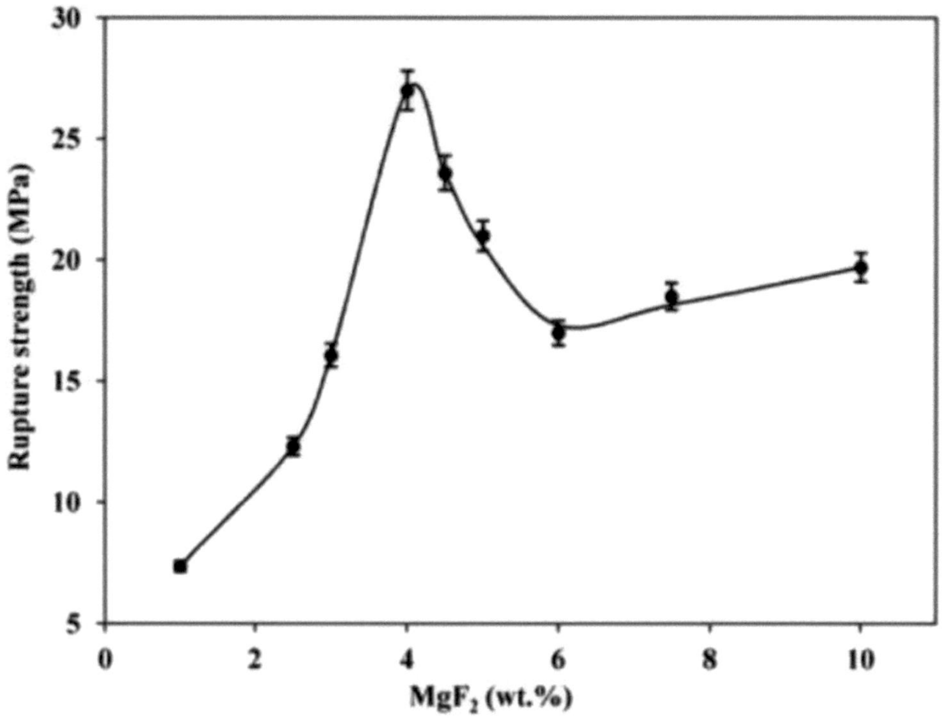

Figure 10. Rupture strength of the TCP-40 wt% TiO_2 composites sintered at 1200°C for 1 h with different percentages of MgF_2.

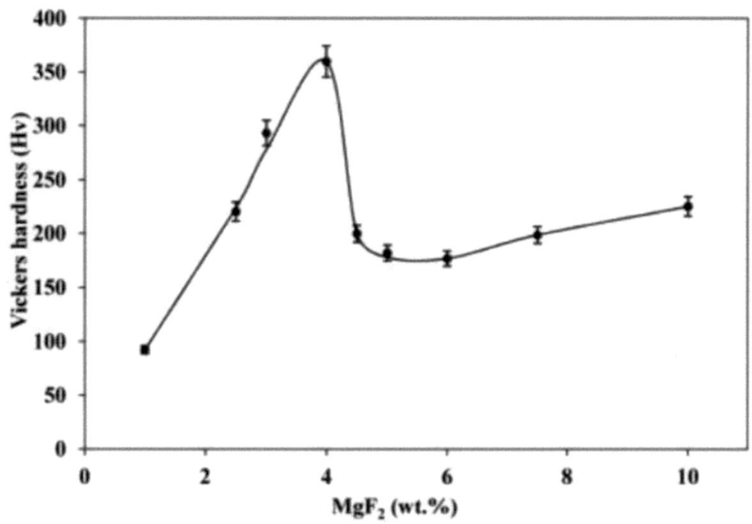

Figure 11. Vickers hardness of the TCP-40 wt% TiO_2 composites sintered at 1200°C for 1 h with different percentages of MgF_2.

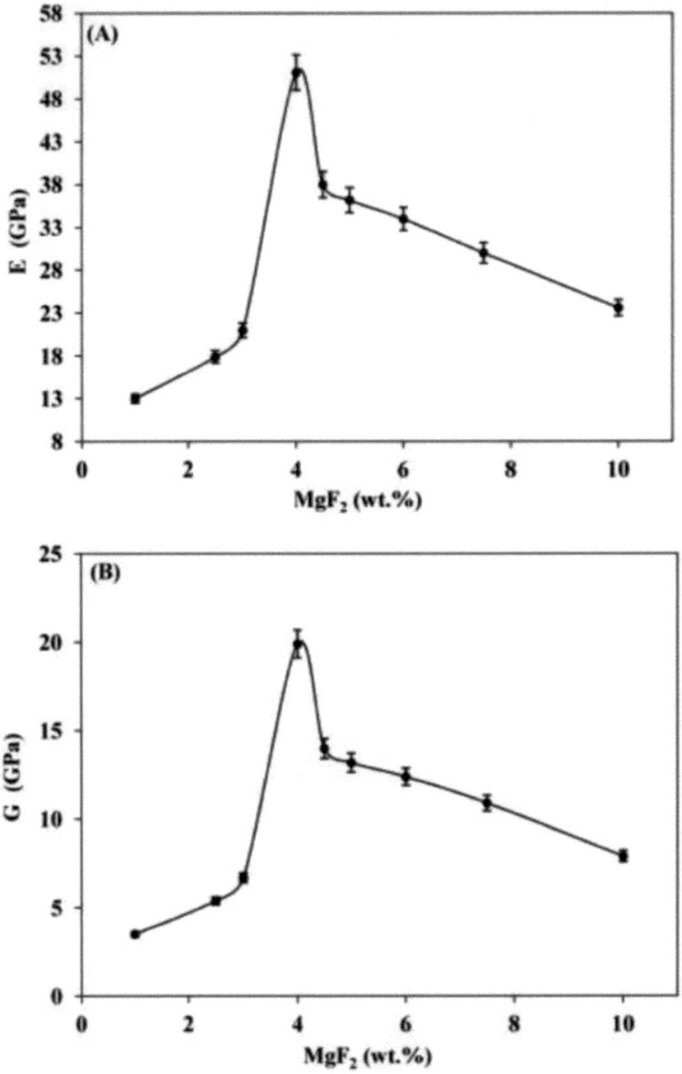

Figure 12. Elastic modulus of the TCP-40 wt% TiO_2 composites sintered at 1200°C for 1 h with different percentages of MgF_2: (A) Young's modulus and (B) the Shear modulus.

Figure 17 shows the XRD patterns of the samples sintered at 1200°C for 1 h without and with different contents of MgF_2 (1, 2.5, 4, 4.5, 5, 7.5 and 10 wt%). The diffraction pattern of the TCP-40 wt% TiO_2 composites sintered without MgF_2 displays the α-TCP phase traces presence (ICDD data file no. 43-1484), the calcium titanate ($CaTiO_3$) (ICDD data file no. 89-8033), the β-TCP phase (ICDD data file no. 70-2065) and the rutile phase of titania (ICDD data file no. 65-1119) (**Figure 17a**). The introduction of 1 wt% MgF_2 to the TCP-40 wt% TiO_2 led to the presence of traces of both calcium titanate ($CaTiO_3$) and fluorapatite (FAp) (ICDD data file no. 76-0558) beside the β-TCP phase and the rutile phase of titania in majority peaks (**Figure 17b**). Further increasing the percentage of MgF_2 to the TCP-TiO_2 composites, we

http://dx.doi.org/10.5772/intechopen.68501

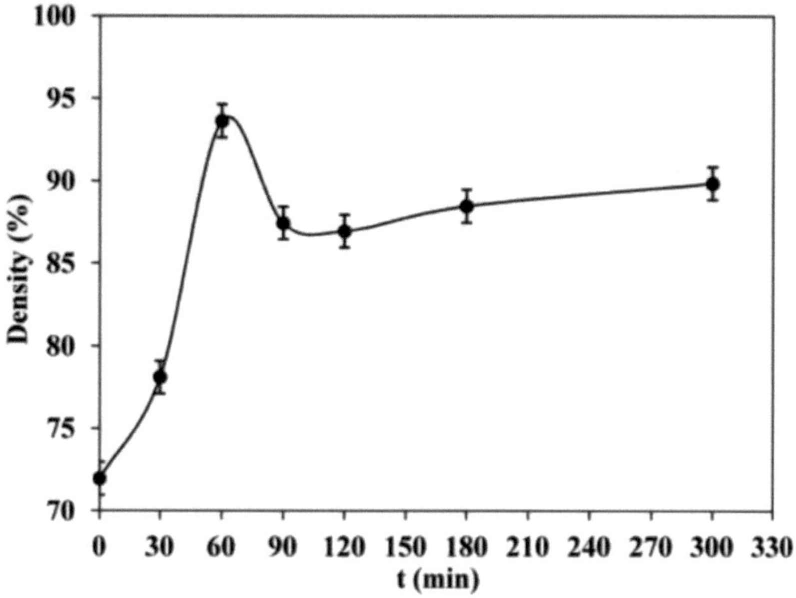

Figure 13. Relative density of the TCP-38 wt% TiO_2-4 wt% MgF_2 composites sintered at 1200°C for different length of the sintering time.

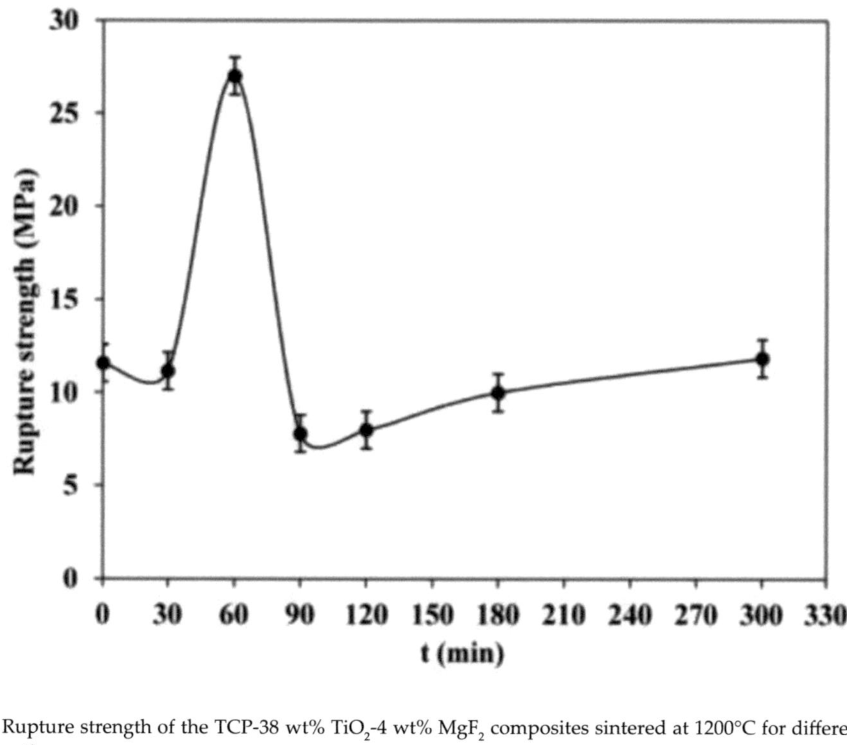

Figure 14. Rupture strength of the TCP-38 wt% TiO_2-4 wt% MgF_2 composites sintered at 1200°C for different length of the sintering time.

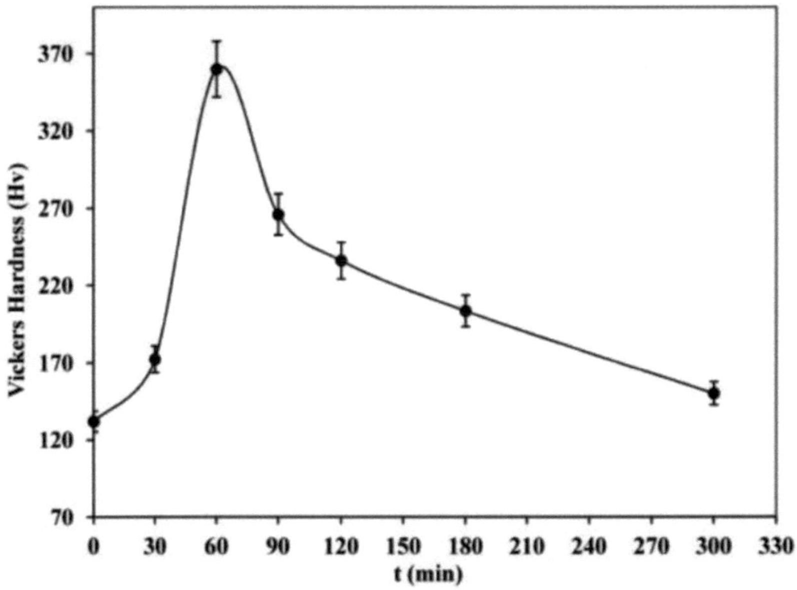

Figure 15. Vickers hardness of the TCP-38 wt% TiO_2-4 wt% MgF_2 composites sintered at 1200°C for different length of the sintering time.

obtain more of fluorapatite (**Figure 17b–h**). Nevertheless, the β-TCP phase is disappeared and the rutile phase quite occupied the majority of peaks (**Figure 17c–h**). The appearance of a new phase relative to $Mg_2(PO_4)F$ (ICDD data file no. 42-0582) was detected after the addition of 2.5, 4 and 4.5 wt% MgF_2 (**Figure 17c–e**). At higher contents of MgF_2 (after 4.5 wt%), the $Mg_2(PO_4)F$ phase is completely disappeared and the intensity of fluorapatite increased (**Figure 17f–h**).

Figure 18 displays the FTIR spectra of the TCP-40 wt% TiO_2 composites sintered at 1200°C for 1 h without and with different percentages of MgF_2 (1, 2.5, 4, 4.5, 5, 7.5 and 10 wt%). Most bands characterize the phosphate group of calcium phosphate (**Figure 18**). Therefore, the bands at 954 and 965 cm^{-1} are assigned to the symmetric stretching of the PO_4^{3} ions while the bands at 1033 and 1093 cm^{-1} are relative to the asymmetric stretching of the PO_4^{3} ions (**Figure 18**).

The ^{31}P MAS-NMR spectra of the TCP-40 wt% TiO_2 composites sintered at 1200°C for 1 h without and with different amounts of MgF_2 (1, 2.5, 4, 4.5, 5, 7.5 and 10 wt%) are illustrated in **Figure 19**. This analysis indicates the presence of the tetrahedral environment, which is assigned to the resonance characteristic of the phosphate group (**Figure 19**). Moreover, the ^{31}P MAS-NMR spectrum of the TCP-40 wt% TiO_2 composites sintered without MgF_2 shows many tetrahedral environments: one broad peak centered at 4 ppm, a shoulder at 0.47 ppm and an intense peak at −0.62 ppm (**Figure 19a**). The NMR spectrum of the TCP-40 wt% TiO_2-1 wt% MgF_2 composites presents an allure modification (**Figure 19b**). One intense peak located at 3 ppm is probably attributed to the phosphor of the phosphate groups of the FAp (**Figure 19c–h**). The second peak at 0.6 ppm is probably relative to the phosphor of the phosphate groups of the

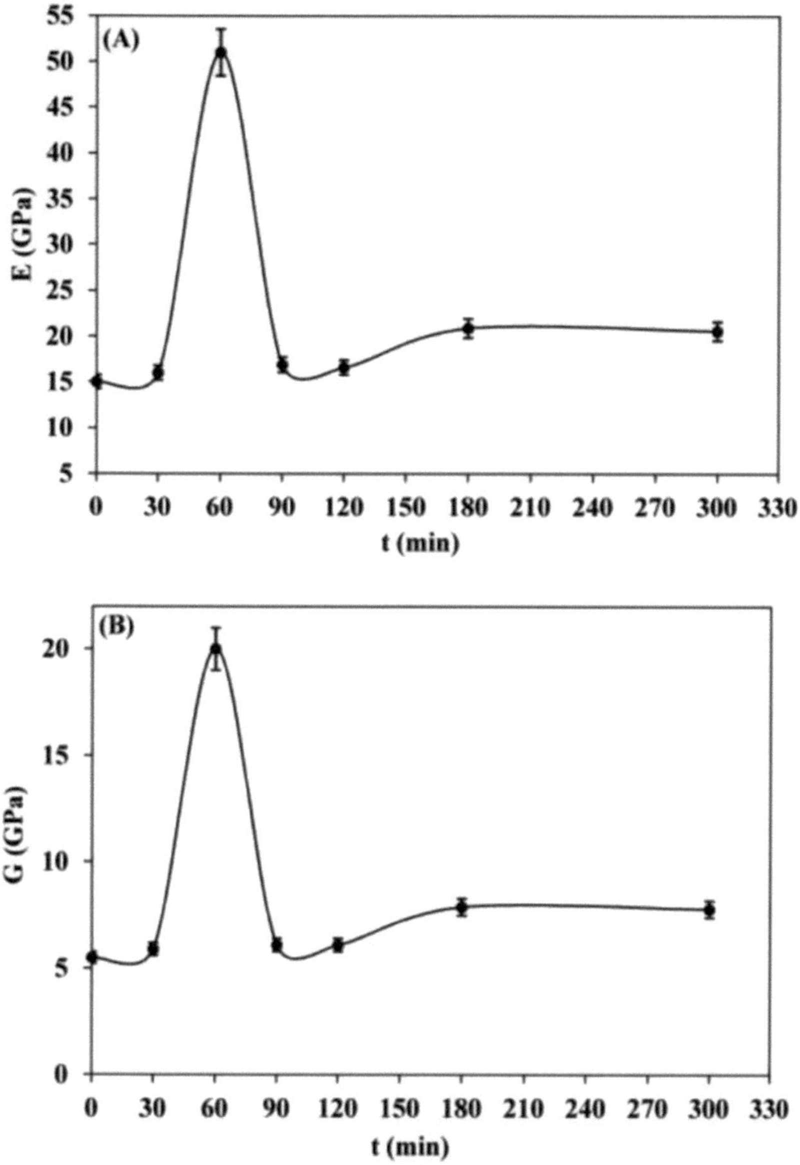

Figure 16. Elastic modulus of the TCP-38 wt% TiO_2-4 wt% MgF_2 composites sintered at 1200°C for different length of the sintering time: (A) Young's modulus and (B) the Shear modulus.

$Mg_2(PO_4)F$ (**Figure 19b–e**). The intensity of the first peak increases with the amount of MgF_2 while the intensity of the second peak decreases and disappears with the TCP-40 wt% TiO_2-4.5 wt% MgF_2 composites (**Figure 19f–h**).

The microstructure of the sintered samples was observed using the SEM analysis. **Figure 20** shows the micrographs of the TCP-40 wt% TiO_2 composites sintered at 1200°C for 1 h

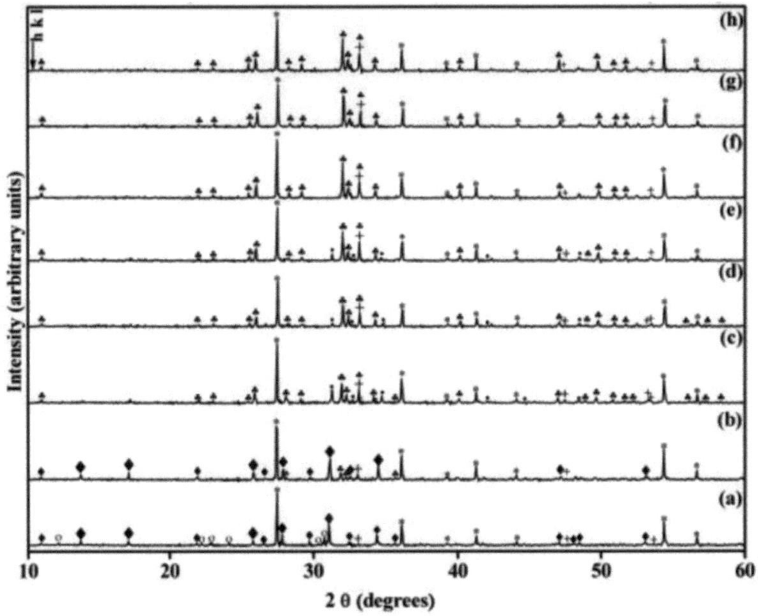

Figure 17. XRD patterns of the TCP-40 wt% TiO_2 composites sintered at 1200°C for 1 h without and with different percentages of MgF_2: (a) 0 wt%, (b) 1 wt%, (c) 2.5 wt%, (d) 4 wt%, (e) 4.5 wt%, (f) 5 wt%, (g) 7.5 wt% and (h) 10 wt%. (o: α-TCP, ♦: β-TCP, +: $CaTiO_3$, *: rutile, ♣: fluorapatite, •: $Mg_2(PO_4)F$).

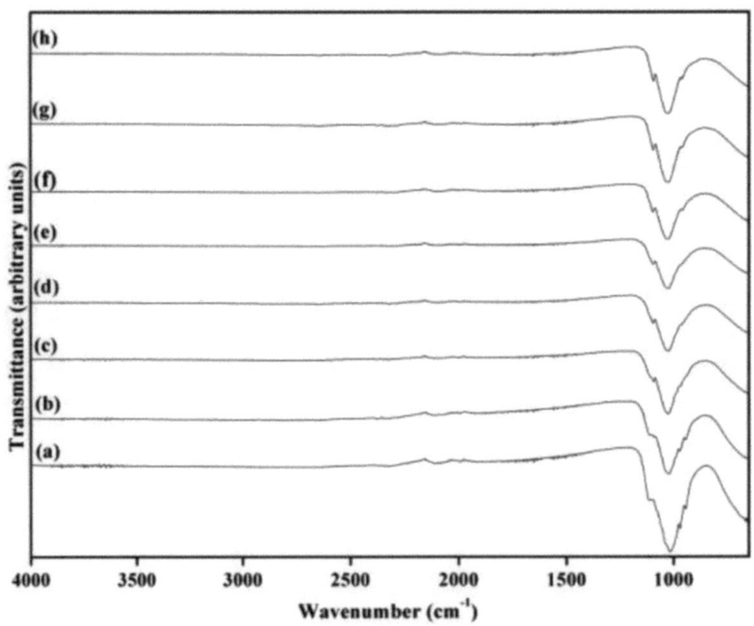

Figure 18. Infrared spectra of the TCP-40 wt% TiO_2 composites sintered at 1200°C for 1 h without and with different percentages of MgF_2: (a) 0 wt%, (b) 1 wt%, (c) 2.5 wt%, (d) 4 wt%, (e) 4.5 wt%, (f) 5 wt%, (g) 7.5 wt% and (h) 10 wt%.

http://dx.doi.org/10.5772/intechopen.68501

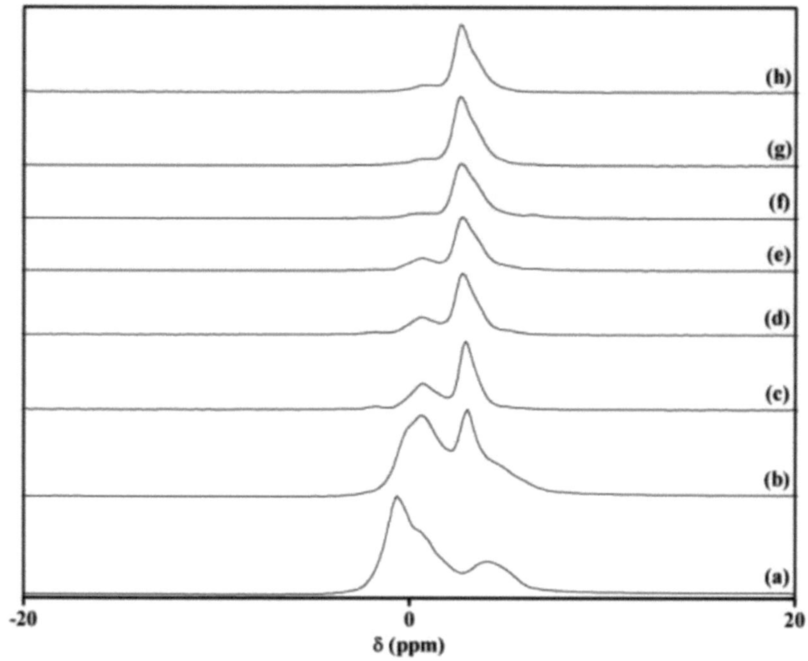

Figure 19. ^{31}P MAS-NMR spectra of the TCP-40 wt% TiO_2 composites sintered at 1200°C for 1 h without and with different percentages of MgF_2: (a) 0 wt%, (b) 1 wt%, (c) 2.5 wt%, (d) 4 wt%, (e) 4.5 wt%, (f) 5 wt%, (g) 7.5 wt% and (h) 10 wt%.

without and with different percentages of MgF_2 (2.5, 4, 7.5 and 10 wt%). The micrograph of the TCP-40 wt% TiO_2 composites sintered without MgF_2 depicts the coalescence of the grains of the TCP and the titania (**Figure 20a**). The microstructure of the TCP-40 wt% TiO_2 composites shows a liquid phase, a continuous phase relative to the β-TCP phases and a small-sized grain relative to the titania (**Figure 20a**). The morphology of the TCP=40 wt% TiO_2 composites sintered with 2.5 wt% MgF_2 was completely transformed (**Figure 20b**). This fact is explained by the intergranular porosity presence beside the grain growth and the formation of a new lamella form (**Figure 20b**). At higher amount of MgF_2 (4 wt%), this new phase still presented in the microstructure (**Figure 20c–d**). An important densification (94%) was registered after the addition of 4 wt% MgF_2 to the TCP-40 wt% TiO_2 composites (**Figure 20c–d**). This result can be explicated by the reduction in the porosity. Moreover, the microstructure of the TCP-38 wt% TiO_2-4 wt% MgF_2 composites shows continuous phases with dense contacts between the grains of the samples (**Figure 20c–d**). Both the addition of 7.5 and 10 wt% of MgF_2 led to the disappearance of the lamella form and the properties of the samples were hindered by the formation in the bubbles form and the grain growth in the microstructure (**Figure 20e–h**). Thus, the microstructure of the TCP-36.25 wt% TiO_2-7.5 wt% MgF_2 composites sintered at 1200°C presents the bubbles form (**Figure 20e–f**). Moreover, the micrographs of the TCP-35 wt% TiO_2-10 wt% MgF_2 depict the grains in hexagonal form formation beside the bubbles form (**Figure 20g–h**).

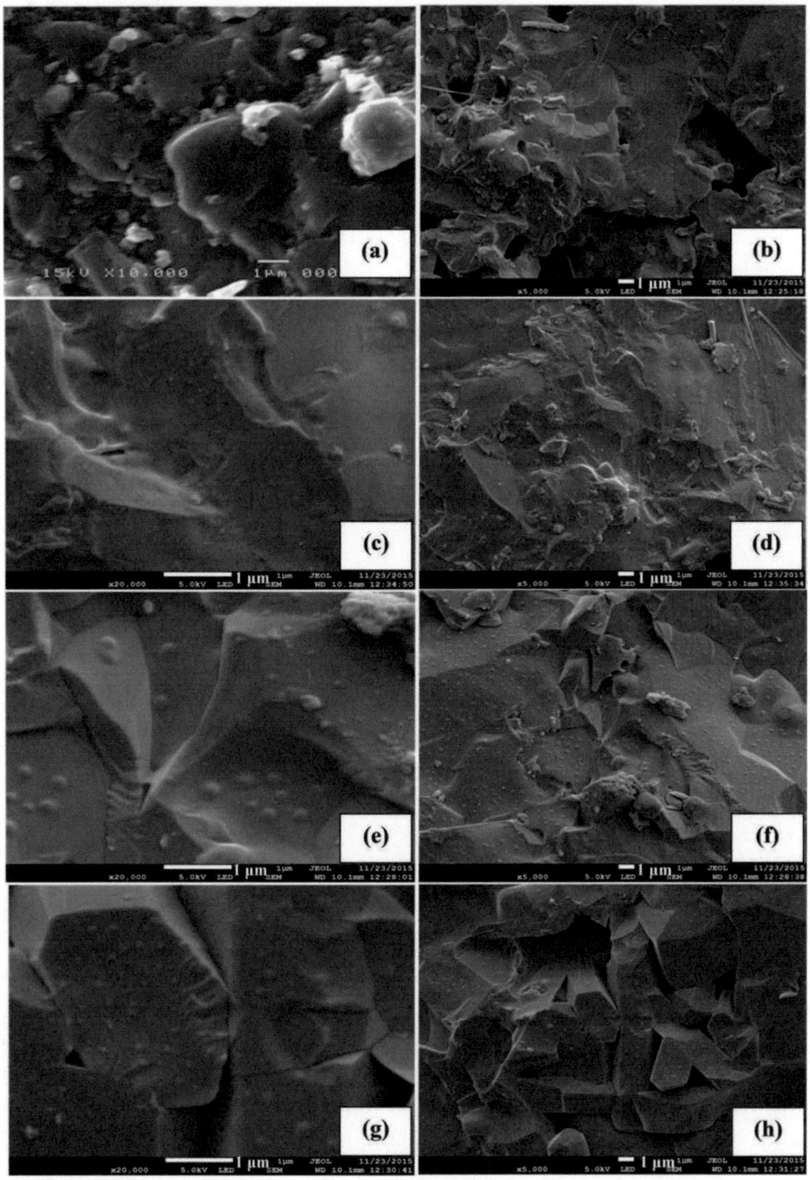

Figure 20. SEM micrographs of the TCP-40 wt% TiO_2 composites sintered at 1200°C for 1 h without and with different percentages of MgF_2: (a) 0 wt% (×10,000), (b) 2.5 wt% (×5000), (c) 4 wt% (×20,000), (d) 4 wt% (×5000), (e) 7.5 wt% (×20,000), (f) 7.5 wt% (×5000), (g) 10 wt% (×20,000) and (h) 10 wt% (×5000).

4. Discussion

In the light of its astonishing properties in terms of biocompatibility, in vivo bioactivity, osteoconductivity and bio-resorbability, tricalcium phosphate has been emerged as one of the most imperative biomaterials [6, 58, 59]. Many authors reported rather fast degradation of the TCP while some researchers observed a minimal or very slow resorption [6, 59, 60]. Consequently, the addition of the inert oxide can enhance the mechanical performances of the TCP in order to combine the high mechanical and tricological properties of titania with the resorbability of TCP. The study of the tricalcium phosphate/titania system interested on improving the densification and the mechanical properties of the TCP by the addition of different amount of TiO_2. The properties of the TCP-TiO_2 composites were increased in accordance with the amount of titania and the sintered temperature. Thus, the optimum values of these composites were acquired after the sintering at 1200°C with 40 wt% of titania. The objective of this study aimed to ameliorate the performances of these composites by the addition of a reinforcement agent like magnesium fluoride (MgF_2). Recently, magnesium, as a dopant in both apatite and β-TCP structures, has been the subject of specific interest due to its important role in biological processes after implantation [61, 62]. Magnesium plays an essential role in the formation of dental caries and bone deposition. Magnesium ions promote bone mineralization and control the growth of the calcium phosphate crystals under in vivo conditions [62, 63]. Magnesium indirectly influences bone resorption and directly stimulates osteoblast proliferation with an effect comparable to that of insulin (a known growth factor for osteoblast) [62–64]. Thus, magnesium containing calcium phosphates is more appropriate for biomaterials and their applications [62]. Magnesium ions play important roles in the formation and transformation of biologically related calcium phosphates [65, 66]. MgF_2 ion is known as the most promising dental caries-promoting element due to a high initial calcification process in the bone formation [37, 38]. Also, fluoride is known to be important in suppressing dental caries [33, 34]. It stimulates the proliferation and the differentiation of bone cells [35]. The fluorine ion (F^-) has been investigated as an essential element for bone and dental formation in the human body [35, 36]. The effect of adding different percentages of MgF_2 on the mechanical properties of the TCP-40 wt% TiO_2 composites sintered at 1200°C for different lengths of the sintering time was investigated. The mechanical performances were studied by Vickers indentation, Brazilian test and ultrasound techniques to determine Vickers hardness (H), the mechanical strength (σ_r) and the elastic moduli (E and G), respectively. The mechanical properties of the TCP-40 wt% TiO_2 composites increase as in terms of the amounts of MgF_2. The optimum values (ρ = 94%, σ_r = 27 MPa, E = 51 GPa, G = 20 GPa and H = 360 Hv) of the TCP-40 wt% TiO_2 composites are attained after the sintering process at 1200°C for 1 h with the addition of 4 wt% MgF_2. These values are relatively similar to those of the bone tissues and precisely the enamel [6, 8, 67–70]. Thus, the enamel is mainly made of the mineral in the calcium phosphate [69, 71]. It is the hardest tissue in the human body because it contains almost no water [69]. For this reason, it is good to get a similar behavior as enamel. In fact, Chun et al. studied the mechanical properties of both enamel and dentin [69].

Therefore, if we compare the value of the strain and stress, enamel tend to fractures earlier than dentin, so it is more brittle than dentine. Nevertheless, if we compare from Vickers hardness value, the enamel is harder than dentine. In fact, dentin was characterized by a higher force resistance while enamel was characterized by a higher wear resistance [69]. The comparison between the previous work derived from the TCP-40 wt% TiO_2 composites and the present work, with adding MgF_2 (the TCP-38 wt% TiO_2-4 wt% MgF_2 composites) leads to the same conclusion described between enamel and dentin. However, as far as hardness is considered, the TCP-38 wt% TiO_2-4 wt% MgF_2 composites are considered harder than the TCP-40 wt%

TiO_2 composites. Such similarity between the present work and the enamel behaviors allows us to study the performances of the TCP-38 wt% TiO_2-4 wt% MgF_2 composites in biomedical applications and specifically as enamel.

The densification discrepancy of the samples prepared by a mixture of the TCP-40 wt% TiO_2 composites with adding 4 wt% MgF_2 (ϱ = 94%) and without MgF_2 (ϱ = 89%) may be due to different phenomena. Thus, the formation of the new compounds and the liquid phase gives rise to the performances of the composites. In general, the very important conditions for the densification by liquid phase sintering are the low viscosity of liquid for fast diffusion of solid through liquid and the solubility of solid to liquid [72] which is responsible for the significant improvement of these properties. Furthermore, several studies have shown the formation of the liquid phase during the sintering process of the tricalcium phosphate/titania system [12, 22]. Thus, Caroff et al. proved that a binary eutectic between the tricalcium phosphate and the titania is relative to this liquid phase [22]. So, in this study, there are two rate-controlling processes in the liquid phase sintering: the dissolution-precipitation phenomena which are dominant sintering mechanism in enhancing the densification by smoothing the contact interface and dissolution of the finer particles and the grain rearrangement occurs due to capillary force between particles. These results are confirmed by literature [72]. So, liquid phase sintering helps to fill the pore in the microstructure of the TCP-TiO_2-MgF_2 composites.

Several examples of the mechanical properties of the bone tissues are shown in **Table 4**. The optimum value of Vickers hardness of the TCP-40 wt% TiO_2 composites was obtained after the addition of 4 wt% MgF_2 (H = 360 Hv). If we compare this result with that of the TCP-40 wt% TiO_2 composites without adding the magnesium fluoride, we notice an increase from 270 to 360 Hv. So, adding 4 wt% MgF_2 enhances Vickers hardness of the TCP-40 wt% TiO_2 composites about 90 Hv. This value (H = 360 Hv) is so close to the value of the enamel (340–370 Hv) (**Table 4**).

Materials	H[a] (Hv)	E[b] (GPa)	G[c] (GPa)	References
Dentin	40–75	18	–	[6, 8, 69]
Enamel	340–370	50–82	–	[8, 69, 70]
Thigh bone	–	20	–	[6, 67, 68]
Cortical bone	–	7–25	–	[6, 67, 68]
β-TCP-40 wt% TiO_2 composites	270	33.1	15.7	Present work
β-TCP-38 wt% TiO_2-4 wt% MgF_2 composites	360	51	20	Present work

[a]Vickers hardness.
[b]Young's modulus.
[c]Shear modulus.

Table 4. Literature examples of the mechanical properties of the bone tissues.

The values of the elastic modulus (E and G) of the TCP-38 wt% TiO_2-4 wt% MgF_2 composites are close to those of the enamel (**Table 4**). In fact, the optimum values of the Shear modulus (G) and Young's modulus (E) of the composites reached 20 and 51 GPa, respectively. If we compare those values with the elastic modulus of the TCP-40 wt% TiO_2 composites sintered without MgF_2, we notice that the elastic corresponding modulus (E and G) of the composites sintered with 4 wt% MgF_2 increase generally about 18 GPa for Young's modulus and 4.3 GPa for the Shear modulus.

The evolution of the mechanical properties (E, G, H and σ_r) of the TCP-38 wt% TiO_2-4 wt% MgF_2 composites with different length of the sintering time is not improved. Thus, the optimum values of the mechanical properties (E = 51 GPa, G = 20 GPa, H = 360 Hv and σ_r = 27 MPa) are obtained after the sintering process for 1 h. In fact, for longer sintering dwell time, these properties are hindered. Indeed, we can retain that the performances (σ_r and H) of the TCP-38 wt% TiO_2-4 wt% MgF_2 composites sintered for 300 min or the TCP-35 wt% TiO_2-10 wt% MgF_2 composites sintered for 60 min are improved which is probably due to the formation of the FAp with more quantities in the TCP-TiO_2-MgF_2 composites. Consequently, the MgF_2 added to the TCP-TiO_2 composites structure enhances their performances with 4 wt%MgF_2.

The diffractograms of the TCP-40 wt% TiO_2 composites sintered at 1200°C with different percentages of MgF_2 indicate that the majority of peaks are relative to the rutile phase of titania (**Figure 17**). The spectrum of the TCP-40 wt% TiO_2 composites sintered without MgF_2 proves the presence of traces of $CaTiO_3$ and α-TCP besides the rutile phase and the β-TCP. The XRD spectra of the TCP-TiO_{2}-MgF_2 composites depicts the formation of both fluorapatite ($Ca_5(PO_4)_3F$) and $Mg_2(PO_4)F$ which is in agreement with the ^{31}P MAS-NMR analysis. Thus, the $Mg_2(PO_4)F$ was obtained only for 2.5, 4 and 4.5 wt% of MgF_2 and disappears for higher amounts of MgF_2 (**Figure 17**).This disappearance is attributed to the incorporation of this phase ($Mg_2(PO_4)F$) into the calcium phosphate structure by a solid state reaction through the compounds. The interaction between TCP and MgF_2 promotes the formation of the $Mg_2(PO_4)F$ and the fluorapatite ($Ca_5(PO_4)_3F$) according to the following reaction (Eq. (7)):

$$2Ca_3(PO_4)_{2(s)} + 2MgF_{2(s)} + TiO_{2(s)} + H_2O \leftrightarrow Ca_5(PO_4)_3F_{(s)} + CaTiO_{3(s)} + Mg_2(PO_4)F_{(s)} + 2HF_{(g)} \quad (7)$$

Several previous analyses (XRD and MAS-NMR) confirm the presence of these compounds [$Mg_2(PO_4)F$ and $Ca_5(PO_4)_3F$].

SEM analyses prove that excellent performances are attained with the TCP-38 wt% TiO_2-4 wt% MgF_2 composites sintered at 1200°C for 1 h. In fact, the continuous phases formed and the new compounds accompanied with the liquid phase appearance improved the performances of the TCP-38 wt% TiO_2-4 wt% MgF_2 composites. These results confirm those obtained with the XRD analysis. Thus, the new lamella form detected in the microstructure of the TCP-38 wt% TiO_2-4 wt% MgF_2 composites is probably accorded to the $Mg_2(PO_4)F$ derived from the solid reaction between MgF2 and TCP according to the previous reaction (Eq. (7)). Above 4 wt% MgF_2, the mechanical properties of the TCP-TiO_2-MgF_2 composites were hindered by the grain growth presence and the bubbles form in the microstructure of composites. Those bubbles resulted from the HF gas evaporation according to the reaction (Eq. (7)). The micrograph of the TCP-35 wt% TiO_2-10 wt% MgF_2 composites indicates the hexagonal grain forms relative to the FAp formation. This phenomenon was accelerated by the dissolution-precipitation phenomena in the liquid phase. Therefore, the liquid phase sintering is used for homogenization and consolidation which presents the advantages of this production method. Thus, the densification of those composites resulted from three processes: the dissolution-precipitation, the rearrangement and the coalescence. This fact is previously confirmed by literature [72–74]. Generally, the microstructure of sintered body by liquid phase sintering mode includes secondary phase, which influences the mechanical, chemical and thermal properties.

5. Conclusion

The effect of the MgF_2 addition on the performances of the TCP-40 wt% TiO_2 composites was studied during the sintering process at 1200°C for different length of the sintering time. The addition of 4 wt% MgF_2 to the TCP-40 wt% TiO_2 composites led to a maximum of both the densification and the mechanical properties after the sintering process at 1200°C for 1 h. A notable amelioration in the elastic modulus and Vickers hardness of the TCP-40 wt% TiO_2 composites was attained with 4 wt% MgF_2 (E = 51 GPa, G = 20 GPa and H = 360 Hv). The enhancement of the composites properties resulted from the formation of new compounds and the liquid phase which permits to fill the pores in the microstructure. Thus, the microstructure indicates the presence of new lamella form relative to the $Mg_2(PO_4)F$ and the hexagonal grains relative to FAp. Above 4 wt% MgF_2, the properties of the composites are hindered by the grain growth formation and the presence of the bubbles form as a result of the dissolution-precipitation phenomena.

Author details

Ibticem Ayadi and Foued Ben Ayed*

*Address all correspondence to: benayedfoued@yahoo.fr

Laboratory of Industrial Chemistry, National Engineering School, Sfax University, Sfax, Tunisia

References

[1] Ben Ayed F, Chaari K, Bouaziz J, Bouzouita K. Frittage du phosphate tricalcique. Comptes Rendus Physique. 2006;7:825. DOI: 10.1016/j.crhy.2006.07.012

[2] Ben Ayed F. Elaboration and characterization of calcium phosphate biomaterial for biomedical application. Rosario Pignatello, editor. In: Handbook of Biomaterials—Physics and Chemistry: In Tech; 2011. pp. 357-374. DOI: 10.5772/23136

[3] Bouslama N, Ben Ayed F, Bouaziz J. Sintering and mechanical properties of tricalcium phosphate-fluorapatite composite. Ceramics International. 2009;**35**:1909-1917. DOI: 10.1016/j.ceramint.2008.10.030

[4] Bouslama N, Chevalier Y, Bouaziz J, Ben Ayed F. Influence of the sintering temperature on Young's modulus and the Shear modulus of tricalcium phosphate-fluorapatite composites evaluated by ultrasound techniques. Materials Chemistry and Physics. 2013;**141**:289-297. DOI: http://dx.doi.org/10.1016/j.matchemphys.2013.05.013

[5] Destainville A, Champion E, Bernache-Assolant D. Synthesis, characterization and thermal behavior of apatitic tricalcium phosphate. Materials Chemistry and Physics. 2003;**80**(1):269-277. DOI: 10.1016/S0254-0584(02)00466-2

[6] Elliott JC. Handbook of Structure and Chemistry of the Apatite and Other Calcium Orthophosphates. Vol. 18. Elsevier Science: BV; 1994. 404 p. ISBN: 9781483290317

[7] Gaasbeek RD, Toonen HG, Van Heerwaarden RJ, Buma P. Mechanism of bone incorporation of β-TCP bone substitute in open wedge tibial osteotomy in patients. Biomaterials. 2005;**26**:6713-6719. DOI: http://dx.doi.org/10.1016/j.biomaterials.2005.04.056

[8] Hench LL. Handbook of an Introduction to Bioceramics, Chapter 37: Characterization of Bioceramics. Hench LL and Wilson J, editors, World Scientific Publishing Co Pte Ltd; Singapore, 1993. pp. 521-540. ISBN: 9789814504164

[9] Hench LL. Bioceramics. Journal of the American Ceramic Society. 1998;**7**:1705-1728. DOI: 10.1111/j.1151-2916.1998.tb02540.x

[10] Jensen SS, Broggini N, Hjorting-Hansen E, Schenk R, Buser D. Bone healing and graft resorption of autograft, anorganic bovine bone and beta-tricalcium phosphate, a histologic and histomorphometric study in the mandibles of minipigs. Clinical Oral Implants Research. 2006;**17**:237-243

[11] Sakka S, Bouaziz J, Ben Ayed F. Mechanical properties of biomaterials based on calcium phosphates and bioinert oxides for applications in biomedicine. Rosario Pignatello, editor. Handbook of Advances in Biomaterials Science and Biomedical Applications. InTech; 2013. pp. 23-50. DOI: 10.5772/53088

[12] Sakka S, Bouaziz J, Ben Ayed F. Sintering and mechanical properties of the alumina-tricalcium phosphate-titania composites. Materials Science and Engineering C. 2014;**40**:92-101. DOI: 10.1016/j.msec.2014.03.036

[13] Sallemi I, Bouaziz J, Ben Ayed F. The effect of adding magnesium oxide on the mechanical properties of the tricalcium phosphate-zirconia composites. Materials Chemistry and Physics. 2015;**151**:50-59. http://dx.doi.org/10.1016/j.matchemphys.2014.11.027 DOI:10.1016/j.matchemphys.2014.11.027#doilink

[14] Sallemi I, Bouaziz J, Ben Ayed F. Elaboration and characterization of bioceramic based on tricalcium phosphate and zirconia. International Journal of Current Engineering and Technology. 2013;**28**:1691-1700. ISSN 2277-4106

[15] Runyan JL, Bennison SJ. Fabrication of flaw-tolerant aluminium titanate-reinforced alumina. Journal of the European Ceramic Society. 1991;**7**:93-99. DOI: 10.1016/0955-2219(91)90006-L DOI:10.1016/0955-2219(91)90006-L#_blank#Persistent link using digital object identifier

[16] Jung S, Kim JH. Sintering characteristics of TiO_2 nanoparticles by microwave processing. Korean Journal of Chemical Engineering. 2010;**27**:645-650. DOI: 10.1007/s11814-010-0057-2

[17] Kaneko H, Uchida M, Kokubo T, Nakamura T. Process of apatite formation induced by anatase on titanium metal in simulated body fluid. In: Brown S, Clarke IC, Williams P, editors. Handbook of Bioceramics 14. Trans Tech Publications Key Engineering Materials; Switzerland, 2002. pp. 649-652. ISBN: 9780878498895

[18] Li D, Chen S, Jing Y, Shao W, Zhang Y, Luan W. The Master sintering curve for pressureless sintering of TiO_2. Science of Sintering. 2007;**39**:103-110. DOI: 10.2298/SOS0702103L

[19] Mazaheri M, Zahedi AM, Haghighatzadeh M, Sadrnezhaad SK. Sintering of titania nanoceramic: densification and grain growth. Ceramics International. 2009;**35**:685-691. DOI: http://dx.doi.org/10.1016/j.ceramint.2008.02.005 DOI:10.1016/j.ceramint. 2008.02.005#doilink

[20] Shin CK, Paek YK. Effect of CuO on the sintering behavior and dielectric characteristics of titanium dioxide. International Journal of Applied Ceramic Technology. 2006;**3**:463-469. DOI: 10.1111/j.1744-7402.2006.02106.x

[21] Ning CQ, Zhou Y. In vitro bioactivity of a biocomposite fabricated from HA and Ti powders by powder metallurgy method. Biomaterials. 2002;**23**:2909-2915. DOI: http://dx.doi.org/10.1016/S0142-9612(01)00419-7 DOI:10.1016/S0142-9612(01)00419-7#doilink

[22] Caroff F, Oh KS, Famery R, Boch P. Sintering of TCP-TiO_2 biocomposites: Influence of secondary phases. Biomaterials. 1998;**19**:1451-1454. DOI: 10.1016/S0142-9612(98)00057-X DOI:10.1016/S0142-9612(98)00057-X#_blank#Persistent link using digital object identifier

[23] Lee JK, Seo DS, Jung HC. Densification and sintering of tricalcium phosphate/titania composite by hot pressing. In: Handbook of the Science of Engineering Ceramics III. T. Ohji, T. Sekino and K. Niihara, editors. Trans Tech Publications Key Engineering Materials; Switzerland, 2006. pp. 101-104. DOI: 10.4028/ www.scientific.net/KEM.317-318.101

[24] Sprio S, Guicciardi S, Dapporto M, Melandri C, Tampieri A. Synthesis and mechanical behavior of β-tricalcium phosphate/titania composites addressed to regeneration of long bone segments. Journal of Mechanical Behavior of Biomedical Materials. 2013;**17**:1-10. DOI: http://dx.doi.org/10.1016/j.jmbbm.2012.07.013 DOI:10.1016/j.jmbbm.2012.07.013#doilink

[25] Manujubala I, Sampath Kumar TS. Preparation of biphasic calcium phosphate doped with magnesium fluoride for osteoporotic applications. Journal of Materials Science Letters. 2001;**20**:1225-1227. DOI: 10.1023/A:1010926923815

[26] Haines J, Leger JM, Gorelli F, Klug DD, Tse JS, Li ZQ. X-ray diffraction and theoretical studies of the high-pressure structures and phase transitions in magnesium fluoride. Physical Review B. 2001;**64**:134110-134131. DOI: https://doi.org/10.1103/PhysRevB.64.134110

[27] Kanchana V, Vaitheeswaran G, Rajagopalan M. High pressure structural phase transitions in magnesium fluoride studied by electronic structure calculations. Journal of Alloys and Compounds. 2003;**352**:60-65. DOI: http://dx.doi.org/10.1016/S0925-8388(02)01158-1 DOI:10.1016/S0925-8388(02)01158-1#doilink

[28] Kusaba K, Kikegawa T. Stable phase with the α-PbO_2 type structure in MgF_2 under high pressure and high temperature. Solid State Communications. 2008;**148**:440-443. DOI: 20081218287

[29] Sun XW, Song T, Liu ZJ, Zhang CR, Tian JH, Guo P. High-pressure and high-temperature bulk modulus of cubic fluorite-type MgF_2 from quasi-harmonic Debye model. Solid State Communications. 2011;**151**:1507-1510. DOI: http://dx.doi.org/10.1016/j.ssc.2011.07.040 DOI:10.1016/j.ssc.2011.07.040#doilink

[30] Babu KR, Lingam CB, Auluck S, Tewari SP, Vaitheeswaran G. Structural, thermodynamic and optical properties of MgF2 studied from first-principles theory. Journal of Solid State Chemistry. 2011;**184**:343-350. DOI: http://dx.doi.org/10.1016/j.jssc.2010.11.025 DOI:10.1016/j.jssc.2010.11.025#doilink

[31] Ozturk H, Kurkcu C, Kurkcu C. High-pressure structural phase transitions and intermediate phases of magnesium fluoride. Journal of Alloys and Compounds. 2014;**597**: 155-160. DOI: http://dx.doi.org/10.1016/j.jallcom.2014.01.221 DOI:10.1016/j.jallcom.2014.01.221#doilink

[32] Ozturk H, Kurkcu C, Kurkcu C. Structural phase transformations and new intermediate phases of MgF_2 under high-pressures applied via conjugate-gradient method. Journal of Alloys and Compounds. 2014;**609**:185-191. DOI: http://dx.doi.org/10.1016/j.jallcom.2014.04.152 DOI:10.1016/j.jallcom.2014.04.152#doilink

[33] Nakade O, Koyama H, Arai J, Ariji H, Takad J, Kaku T. Stimulation by low concentrations of fluoride of the proliferation and alkaline phosphatase activity of human dental pulp cells in vitro. Archives of Oral Biology. 1999;**44**:89-92. DOI: 10.1016/S0003-9969(98) 00099-5

[34] Evis Z, Pinar Sun Z. Structural and mechanical investigations of magnesium and fluoride doped nanosize calcium phosphates. Journal of Ceramic Processing Research. 2010;**11**(6):701-715

[35] Farley JR, Wergedal JE, Baylink DJ. Fluoride directly stimulates proliferation and alkaline phosphatase activity of bone-forming cells. Science. 1983;**222**(4621):330-332. DOI: PMID:6623079

[36] Fathi MH, Zahrani EM. The effect of rotation speed and time of milling on synthesis and properties of fluoridated hydroxyapatite biomaterial. Iranian Journal of Pharmaceutical Science. 2008;**4**(3):201-208. ISSN: 1735-2444

[37] Kim SH, Lee HC, Bang HG, Park SY. Effect of MgF_2 additive on the mechanical properties in Hydroxyapatite/Zirconia composites. In: Handbook of Eco-Materials Processing & Design VII. Hyung Sun Kim, Yu Bao Li and Soo Wohn Lee, editors. Trans Tech Publications Materials Science Forum; Switzerland, 2006. pp. 478-481. DOI: 10.4028/www.scientific.net/MSF.510-511.478

[38] Gu XN, Zheng YF. A review on magnesium alloys as biodegradable materials. Frontiers of Materials Science in China. 2010;**4**(2):111-115. DOI: 10.1007/s11706-010-0024-1

[39] Xue W, Dahlquist K, Banerjee A, Bandyopadhya A, Bose S. Synthesis and characterization of tricalcium phosphate with Zn and Mg based dopants. Journal of Materials Science: Materials in Medicine. 2008;**19**:2669-2677. DOI: 10.1007/s10856-008-3395-4

[40] Chang CS, Hon MH. Texture effect of hot-pressed magnesium fluoride on optical transmittance. Materials Chemistry and Physics. 2003;**81**(1):27-32. DOI: http://dx.doi.org/10.1016/S0254-0584(02)00460-1

[41] Nofar M, Madaah Hosseini HR, Shivaee HA. The dependency of optical properties on density for hot pressed MgF_2. Infrared Physics and Technology. 2008;**51**:546-549. DOI: http://dx.doi.org/10.1016/j.infrared.2008.06.002 DOI:10.1016/j.infrared.2008.06.002#doilink

[42] Evis Z, Usta M, Kutbay I. Improvement in sinterability and phase stability of hydroxyapatite and partially stabilized zirconia composites. Journal of the European Ceramic Society. 2009;**29**:621-628. DOI: http://dx.doi.org/10.1016/j.jeurceramsoc.2008.07.020 DOI: 10.1016/j.jeurceramsoc.2008.07.020#doilink

[43] Pardun K, Treccani L, Volkmann E, Streckbein P, Heiss C, Gerlach JW, Maendl S, Rezwan K. Magnesium containing mixed coatings on zirconia for dental implants: Mechanical characterization and in vitro behavior. Journal of Biomaterials Applications. 2015;**30**(1):104-118. DOI: 10.1177/0885328215572428

[44] Chen X, Chen X, Brauer DS, Wilson RM, Hill RG, Karpukhina N. Novel alkali free bioactive fluorapatite glass ceramics. Journal of Non-Crystalline Solids. 2014;**402**:172-177. DOI: 10.1016/j.jnoncrysol.2014.05.025

[45] Kanazawa T, Umegaki T, Yamashita K, Monma H, Hiramatsu T. Effects of additives on sintering and some properties of calcium phosphates with various Ca/P Ratios. Journal of Materials Science. 1991;**26**:417-422. DOI: 10.1007/BF00576536

[46] Hench LL. Bioceramics: From concept to clinic. Journal of the American Ceramic Society. 1991;**74**:1487-1510. DOI: 10.1111/j.1151-2916.1991.tb07132.x

[47] Azami M, Jalilifiroozinezhad S, Mozafari M, Rabiee M. Synthesis and solubility of calcium fluoride/hydroxy-fluorapatite nanocrystals for dental applications. Ceramics International. 2011;**37**:2007-2014. DOI: http://dx.doi.org/10.1016/j.ceramint.2011.02.025 DOI:10.1016/j.ceramint.2011.02.025#doilink

[48] Carsten J, Marco B. ^{19}F NMR spectroscopy of glass ceramics containing fluorapatites. Biomaterials.1996;**17**(21):2065-2069. PMID: 8902239

[49] Sakka S, Ben Ayed F, Bouaziz J. Mechanical properties of tricalcium phosphate alumina composites. Cheikhrouhou A, editor. In: Proceedings of the IOP Conference Series on Materials Science and Engineering. Mahdia Tunisia, 2012;**28**:012028. DOI:10.1088/1757-899X/28/1/012028

[50] Brunauer S, Emmet PH, Teller J. Adsorption of gases in multimolecular layers. Journal of the American Chemical Society. 1938;**60**:310-319. DOI: 10.1021/ja01269a023

[51] Bernache-Assollant D. Handbook of Chimie-physique du Frittage Forceram Formation Céramique. 1st ed. Lavoisier: Hermès Science Publications; 1993. 345 p. ISBN: 9782866013431

[52] ASTM C496. Standard test method for splitting tensile strength of cylindrical concrete specimens. In: Annual Book of ASTM Standards. Philadelphia: ASTM; 1984. p. 336. DOI: 10.1520/C0496_C0496M-11

[53] ISRM. Suggested methods for determining tensile strength of rock materials. International Journal of Rock Mechanics and Mining Sciences and Geomechanics Abstracts. 1978;**15**:99-103. DOI: 10.1016/0148-9062(78)91677-7 DOI:10.1016/0148-9062(78)91677-7#_blank#Persistent link using digital object identifier

[54] ASTM C1327-96. Standard test method for Vickers indentation hardness of advanced ceramics. In: Annual Book of ASTM Standards. West Conshohocken, United States, 2003

[55] Chevalier Y. Essais dynamiques sur composites—caractérisation aux hautes fréquences. Techniques de l'Ingénieur. 2003. pp. 1-19. REF:AM5401 V1

[56] Bouslama N, Ben Ayed F, Bouaziz J. Effect of fluorapatite additive on densification and mechanical properties of tricalcium phosphate. Journal of the Mechanical Behavior of Biomedical Materials. 2010;**3**:2-13. DOI: http://dx.doi.org/10.1016/j.jmbbm.2009.01.007 DOI: 10.1016/j.jmbbm.2009.01.007#doilink

[57] Gouma PI, Mills MJ. Anatase to rutile transformation in titania powders. Journal of the American Ceramic Society. 2001;**84**:619-622. DOI: 10.1111/j.1151-2916.2001.tb00709.x

[58] Vallet-Regi M, Gonzalez-Calbet JM. Calcium phosphates as substitution of bone tissues. Progress in Solid State Chemistry. 2004;**32**:1-31. DOI: http://dx.doi.org/10.1016/j.progsolidstchem.2004.07.001 DOI: 10.1016/j.progsolidstchem.2004.07.001#doilink

[59] Williams D. An introduction to medical and dental materials. In: Williams D, editor. Concise Encyclopedia of Medical and Dental Materials. Pergamon Press and the MIT Press. Pergamon, 1990. pp. xvii-xx

[60] LeGeros RZ, LeGeros JP, Daculsi G, Kijkowaka R. Calcium phosphate biomaterials: Preparation, properties, and biodegradation. In: Wise DL, Trantolo DJ, Altobelli DE, et al., editors. Encyclopedic Handbook of Biomaterials and Bioengineering, Part A: Materials. New York: Marcel Dekker; 1995. pp. 1429-1463

[61] Kannan S, Neunhoeffer FG, Neubauer J, Rebelo AHS, Valério P, Ferreira JMF. Rietveld structure and in vitro analysis on the influence of magnesium in biphasic hydroxyapatite and β-tricalcium phosphate mixtures. Journal of Biomedical Materials Research Part B: Applied Biomaterials. 2009;**90**(1):1552-4981. DOI: 10.1002/jbm.b.31299

[62] Gozalian A, Behnamghader A, Daliri M, Moshkforoush A. Synthesis and thermal behavior of Mg-doped calcium phosphate nanopowders via the sol gel method. Scientia Iranica Transactions F: Nanotechnology. 2011;**18**(6):1614-1622. DOI: 10.1016/j.scient.2011.11.014

[63] Serre CM, Papillard M, Chavassieux P, Voegel JC, Boivin G. Influence of magnesium substitution on a collagen-apatite biomaterial on the production of a calcifying matrix by human osteoblasts. Journal of Biomedical Materials Research. 1998;**42**:626-633. DOI: 10.1002/(SICI)1097-4636(19981215)42:4<626::AID-JBM20>3.0.CO;2-S

[64] Suchanek WL, Byrappa K, Shuk P, Riman RE, Janas VF, TenHuisen KS. Mechanochemical-hydrothermal synthesis of calcium phosphate powders with coupled magnesium and carbonate substitution. Journal of Solid State Chemistry. 2004;**177**:793-799. DOI: http://dx.doi.org/10.1016/j.jssc.2003.09.012 DOI:10.1016/j.jssc.2003.09.012#doilink

[65] LeGeros RZ. Calcium phosphates in oral biology and medicine. In: Myers HM, editor. Handbook of Monographs in Oral Sciences.1st ed. Basel: Karger; 1991. pp. 1-200. PMID: 1870604

[66] Otsuka M, Oshinbe A, LeGeros RZ, Tokudome Y, Ito A, Otsuka K, Higuchi WI. Efficacy of the injectable calcium phosphate ceramics suspensions containing magnesium, zinc and fluoride on the bone mineral deficiency in ovariectomized rats. Journal of Pharmaceutical Sciences. 2008;**97**(1):421-432. DOI: 10.1002/jps.21131

[67] Ravaglioli A, Krajewski A. Handbook of Bioceramics: Materials, Properties and Appli cations. London, England: Chapman & Hall; 1992. 422 p. DOI: 10.1007/97 8-94-011-2 336-5

[68] Shi D, editor. Handbook of Biomaterials and Tissue Engineering. 1st ed. New York: Springer-Verlag Berlin Heidelberg; 2004. 200 p. DOI: 10.1007/978-3-662-06104-6

[69] Chun KJ, Choi HH, Lee JY. Comparison of mechanical property and role between enamel and dentin in the human teeth. Journal of Dental Biomechanics. 2014;**5**:1-6. DOI: 10.1177/1758736014520809 DOI:10.1177%2F1758736014520809#pmc_ext

[70] Biswas N, Dey A, Kundu S, Chakraborty H, Mukhopadhyay AK. Mechanical properties of enamel nanocomposite. ISRN Biomaterials Hindawi. 2013;**2013**:1-15. DOI: http://dx.doi.org/10.5402/2013/253761

[71] Staines M, Robinson WH, Hood JAA. Spherical indentation of tooth enamel. Journal of Materials Science. 1981;**16**(9):2551-2556. DOI: 10.1007/BF01113595

[72] Hosokawa M, Nogi K, Naito M, Yokoyama T, editors. Handbook of Nanoparticle Technology. 1st ed. Linacre House, Amsterdam, The Netherlands: Elsevier BV; 2007. 644 p. ISBN: 9780080558028

[73] Ben Ayed F, Bouaziz J, Bouzouita K. Pressureless sintering of Fluorapatite under oxygen atmosphere. Journal of the European Ceramic Society. 2000;**20**(8):1069-1076. DOI: 10.1016/S0955-2219(99)00272-1

[74] Ben Ayed F, Bouaziz J, Bouzouita K. Calcination and sintering of fluorapatite under argon atmosphere. Journal of Alloys and Compounds. 2001;**322**(1-2):238-245. DOI: http://dx.doi.org/10.1016/S0925-8388(01)01200-2 DOI:10.1016/S0925-8388(01)01200-2#doilink

Chapter 4

Evolution of Magnetic Properties in Ferrites: Trends of Single-Sample and Multi-Sample Sintering

Ismayadi Ismail, Idza Riati Ibrahim and
Rodziah Nazlan

Additional information is available at the end of the chapter

http://dx.doi.org/10.5772/intechopen.68500

Abstract

Microstructure of magnetic materials greatly influences the performance of magnetic properties, and sintering has been used as an agent to tailor the microstructure of these magnetic materials especially ferrites. Nanostructured ferrites prepared by high-energy milling method are often inherently unstable owing to their small constituent sizes, non-equilibrium cation distribution, disordered spin configuration, and high chemical activity. Therefore, sintering of the milled ferrites recrystallizes the nanostructure and causes its transition from an excited metastable (activated) state into the low-energy crystalline state. A better understanding of the response of nanoscale ferrites with changes in temperature is crucial not only for basic science (the development of an atomistic and microscopic theory of the mechanochemical processes) but also because of the technological high-temperature applications in catalysis, ferrofluids and information storage. This chapter discusses on two different sintering schemes, which are a commonly applied multi-sample sintering and a rarely adopted single-sample sintering. Experimental results of single-sample and multi-sample sintering of NiZn ferrites and yttrium iron garnet (YIG) were highlighted, and their microstructural consequences on the magnetic properties were also discussed.

Keywords: microstructural evolution, BH-hysteresis, ferrites, NiZn ferrites, yttrium iron garnet (YIG)

1. Sintering as a microstructure tailoring agent

Microstructure of polycrystalline ceramics is usually complex, consisting of grains, grain boundaries, porosity and secondary phases. This kind of structure is not seen in single crystals. Variations in the microstructure with different kinds of shape, size, distribution and orientation of the grains play a key role in many of the macroscopic properties including magnetic, thermophysical, mechanical, electrical and many other properties. Essentially, these phenomena are familiar with the polycrystalline ceramics samples having micronic grain size, and the information on their relationships is well understood. Materials in the micrometer scale mostly exhibit physical properties the same as that of bulk form; however, materials in the nanometer scale may exhibit physical properties distinctively different from that of micrometer scale. Nanomaterials may have significantly lower melting point or phase transition temperature and appreciably reduced lattice constants, due to a huge fraction of surface atoms in the total amount of atoms [1]. Materials with an altered 'nano'-microstructure provide potential for new or improved applications [2]. Sintering has been known as an agent to alter the microstructure condition of a polycrystalline material. Through optimization of sintering conditions such as sintering temperature, sintering atmosphere, heating and cooling rates, sintering time and partial pressure of sintering atmosphere, the best materials properties could be achieved. Tailoring the microstructure to attain certain desirable materials properties is the main challenge and of interest in material science.

2. Single-sample and multi-sample sintering

There are two different sintering schemes in producing polycrystalline materials, which are commonly applied multi-sample sintering scheme and rarely adopted single-sample sintering scheme. Generally, reported studies involving sintering and materials properties employed multi-sample sintering scheme [3–9]. The multi-sample sintering has as many starting compacts as the number of the intended sintering temperatures where each sample sintered only once at different temperatures. All compacts are assumed to have identical morphologies, for example, particle size distribution. However, a rarely adopted single-sample sintering scheme has only one single compact with definite starting point and one particular particle size distribution where only one sample sintered at different temperatures. Therefore, multi-sample sintering is subjected to possible statistical errors since the particle size distributions for all the samples may not be as identical as assumed as compared to that of single-sample sintering. Thus, more convincing data could be obtained for the scientific interpretation of the evolution study. Besides, it is more economical with respect to raw materials and sample preparation time. The schematic of the different sintering schemes is been shown in **Figure 1**. Ceramic is defined as the art and science of making and using solid articles [10]. The statement made a clear view that it is not an easy task to produce the same ceramic with almost the same properties because ceramic is composed of a complex system. The question is whether a material scientist is able to make such ceramic by employing single-sample and multi-sample sintering, with almost the same or enhanced properties is of great interest.

http://dx.doi.org/10.5772/intechopen.68500

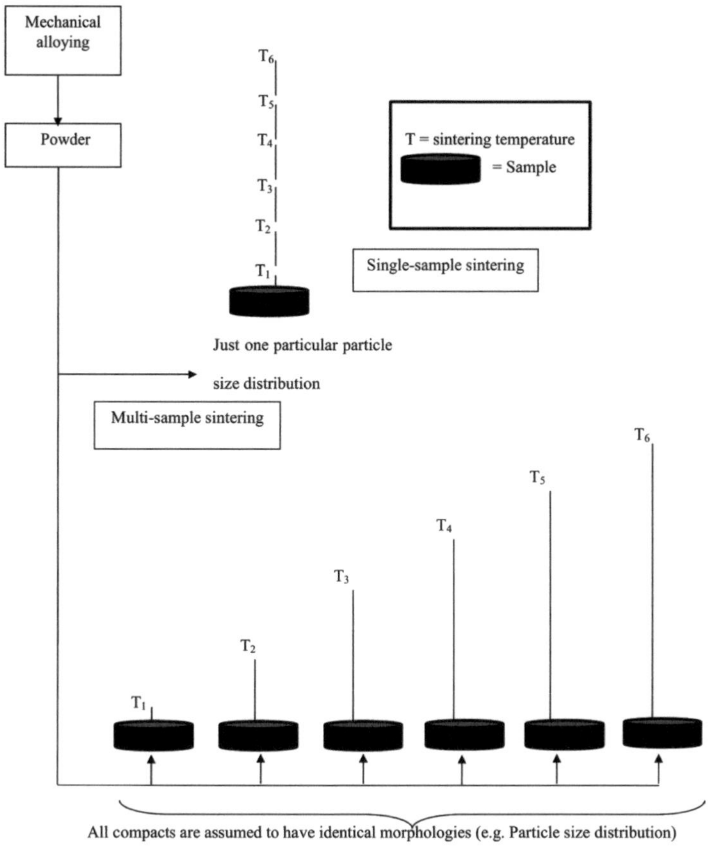

Figure 1. Schematic of different sintering schemes: multi-sample and Single-sample sintering.

3. Magnetic properties evolution and the research gap

Evolution in magnetic properties is laterally correlated with the evolution of the microstructure, particularly from nanometric to micronic regime of grains as shown in **Figure 2**. However, the reported cause and effect sequences, in the magnetic properties research literature, are an experimental sequence focused mainly on yielding the final outcome, for example, the final microstructure-magnetic properties relationship at final sintering temperature. Therefore, microstructural dependence of magnetic properties for polycrystalline ferrite having micrometer grain size has been widely studied and greatly understood. However, research on ferrite from the nanometer scale has been a field of intense study, due to the novel properties shown by particles located in the transition region between the isolated atoms and bulk solids. Their novel properties make them attractive, both from the scientific knowledge of understanding

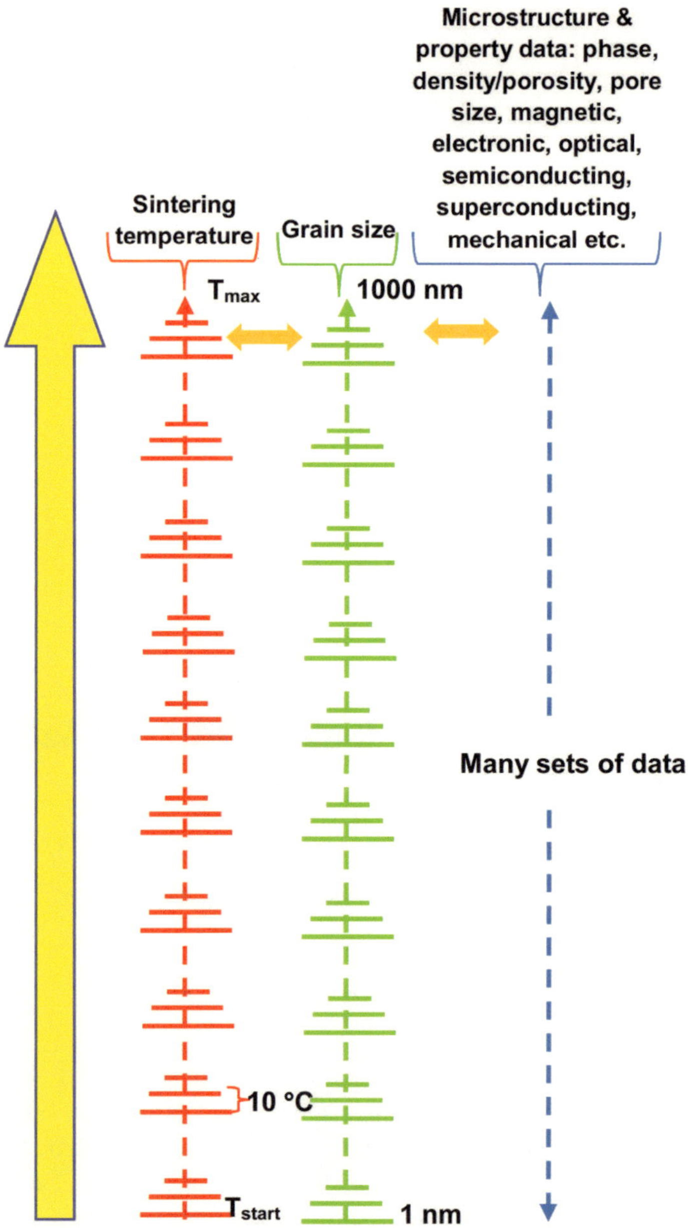

Figure 2. Grand evolution-data acquisition scheme for polycrystalline materials.

their properties, and the technological importance of enhancing the performance of the present materials. A fundamental line of scientific enquiry thus has been neglected, particularly by ferrite researchers for more than 70 years: What would be the magnetic-microstructure relationships at various intermediate sintering conditions during the parallel evolutions of the morphology and magnetic properties? Therefore, much possible essential development information has been neglected, thus reducing the capabilities of producing good fundamental scientific knowledge, which lies behind the parallel evolution of the microstructure-material properties, particularly in magnetic properties. This absence information has leaving behind many research gaps and research questions that have to be solved in this study:

i. How microstructural properties evolve with the magnetic properties from several nanometers to micrometer grains size?

ii. How the evolution of an amorphous-crystalline mixture state to complete polycrystalline state affects the microstructure-magnetic properties?

iii. Does sample with nanometer grain size demonstrates the similar magnetic properties with samples having micron grain size?

iv. What is the relationship of evolving microstructure properties with magnetic properties of material?

v. When is the remarkable transition of magnetic properties between the nanometer and micrometer grain size?

Hence, revealing the systematic development of grains having size from several nanometers up to micrometer is an important parameter of designing best materials properties. The evolution has yet to be established in wide variation of properties since the knowledge of parallel evolution of microstructure and properties is absence in these materials:

i. Other electroceramics, for example, high-temperature heating elements and electrodes, voltage dependent resistors, thermally sensitive resistors, solid fast-ion conductors, humidity and gas sensors, superconductors

ii. Thermo-mechanical ceramics, for example, SiC, Al_2O_3

iii. Rare-earths-based magnetic materials, for example, $Nd_9Fe_{14}B$, $SmCo_{17}$

iv. Transition metal-alloys based magnetic materials, for example, NiFe, NiCo

v. Mechanical metals and metallic alloys (elemental), for example, pure iron, copper, chromium, nickel/carbon steels

vi. Polycrystalline semiconductors, for example, zinc oxide

4. Introduction to ferrites

Generally, a class of magnetic oxide, which contains iron oxide as a primary component, is commonly described as ferrites. Similar to most ceramic material, the physical properties of ferrite are also hard but brittle. The drastic progress in the development and growth of ferrites for technological application has once force the industry to leave behind the research and study in ferrite. The industrial importance of ferrites becomes apparent when one examines the diversity of their applications. Ferrite has been extensively used in various electronic devices. These applications include choke filters [11], transformers [12], antenna rods [13], microwave devices [14, 15], isolators [16], circulators [17], phase shifters [18] and many others. The frequencies of the applications range from direct current (DC) to the highest one at which any electronic device can function [19].

Ferrite exhibits ferrimagnetic behavior which possesses unequal, anti-parallel ionic magnetic moments resulting in a net moment due to incomplete compensation. There are three classes of commercial ferrite in the industry, and each of the types has their own specific crystal structure. The three classes of the commercial ferrites are as follows:

i. Soft ferrite with spinel cubic structure, for example, nickel zinc ferrite and manganese zinc ferrite.

ii. Soft ferrite with garnet structure, for example, yttrium-based garnets that are used in microwave devices.

iii. Hard ferrite with magnetoplumbite structure, for example, barium hexaferrite and strontium hexaferrite. The hexagonal ferrites develop high coercivity and are an important member of the permanent magnet family.

5. Sintering of ferrite materials

Ferrites are commonly fabricated via two major techniques: the conventional technique and the non-conventional technique. Through the conventional ceramic technique, the raw material powders are mixed and sintered at over 1000°C sintering temperature. This process allows interdiffusion of atoms in a pre-selected composition to form a mixed crystal. The other technique for preparing ferrites is the non-conventional technique. A non-conventional technique in a liquid medium may produce intermediate, finely divided mixed hydroxides or mixed organic salts, which would assist the subsequent diffusion process [20].

Various synthesis methods including the conventional and non-conventional techniques of ferrites preparation have been shown in **Table 1**. The table shows that the formation of ferrite through non-conventional technique could be produced by using the obtained fine powders at much lower sintering temperature. Yet most of the techniques still require sintering, although at relatively lower temperatures to produce a single phase material. The sintering temperature could be as low as 200°C [21], though displaying least performance of magnetic properties compare to much higher sintering temperatures. Highest sintering temperature is normally employed for synthesizing bulk ferrites via solid state reaction and has been shown to produce optimum magnetic properties (see **Table 1**). However, low sintering temperature is required for nano-sized materials as basic requirement since magnetic properties of the bulk materials differ drastically from the nano-sized materials.

Sintering in certain condition of atmosphere would form pure phase as has been observed in reference [22]. Furthermore, sintering atmosphere is also responsible in altering the magnetic properties and the effects can be observed in cation redistribution and oxygen deficiency [21]. The sintering time displays in **Table 1**, which shows that the non-conventional technique requires much shorter sintering time as compare to conventional technique. The right selection of sintering times would result in high densification and homogeneous materials which are largely important in magnetic materials since less densified materials result in hindrance to the domain wall movement, thus reducing the total magnetization. In addition, the

http://dx.doi.org/10.5772/intechopen.68500

Material	Synthesis method	Starting materials	Sintering conditions	Microstructure features	Optimum magnetic properties	Reference
$Ni_{0.3}Cu_{0.2}Zn_{0.5}Fe_2O_4$	Citrate precursor method	$Fe(NO_3)_3.9H_2O$ $Zn(NO_3)_2.\ 4H_2O$ $Cu\ (NO_3)_2.3H_2O$ $Ni(NO_3)_2.6H_2O$ Citric acid	Sintering atmosphere: air-, argon-, carbon monoxide-ambient atmospheres Sintering temperature: 200,400 and 600°C Sintering time: 2 h	Crystallite sizes increased with increased sintering temperatures for all sintering atmosphere ranging from 22.7 to 28.1 nm for air atmosphere, 8.8–24. 7 nm for argon atmosphere and 10.1–31.6 nm for carbon monoxide-ambient atmosphere	Saturation magnetization, M_s: 69.1 emu/g for sample sintered at 600°C under carbon monoxide-ambient atmosphere	[21]
$NiFe_2O_4$	Sol-gel auto-combustion method	$Fe(NO_3)_3{\cdot}9H_2O$ $Ni(NO_3)_2{\cdot}6H_2O$ Citric acid	Sintering temperature: 900, 1100, 1200 and 1300°C Sintering time: 2 h	Average grain sizes increased from 0.1 to 2.2 μm with increasing sintering temperatures Density increased from 2.93 to 4.30 g/cm^3 with increasing sintering temperatures	M_s: ~50.0 emu/g for sample sintered at 1100°C	[25]
$Ni_{0.5}Zn_{0.5}Fe_2O_4$	Co-precipitation method	$NiCl_2$ $ZnCl_2$ $FeCl_3$ NaOH	Sintering temperature: 800 and 900°C Sintering time: 3 h	Average crystallite sizes increased from 38.4 to 42.2 nm with increasing sintering temperatures	M_s: 89.5 emu/g for sample sintered at 900°C	[26]
$Ni_{0.266}Zn_{0.66}Cu_{0.09}Fe_{1.968}O_{4-\delta}$	Solid-state reaction method	NiO ZnO CuO Fe_2O_3	Sintering atmosphere: Air Sintering temperature: 1050°C Sintering time: 1, 3, 5 and 7 h	Average crystallite sizes increased from 3.9 to 12.9 μm with increasing sintering times Density increased from 5.06 to 5.21 g/cm^3 with increasing sintering times	M_s: 245.1 emu/cm^3 for sample sintered for 7 h Initial permeability, μ_i, at 1 kHz: 2305 for sample sintered for 7 h	[23]

Material	Synthesis method	Starting materials	Sintering conditions	Microstructure features	Optimum magnetic properties	Reference
$Y_3Fe_5O_{12}$	Solid-state reaction method	Fe_2O_3 Y_2O_3	Two different sintering curves Curve 1 Sintering temperatures, T_1:1200–1450°C Sintering time: none Heating rate:10°C/min Cooling rate: 2°C/min Curve 2 Sintering temperatures, T_1:1350°C, T_2:1200 °C T_1:1350°C, T_2:1300 °C T_1:1450 °C, T_2:1300 °C Sintering time at T_2: 6, 12, 18 and 24 h Heating rate:10°C/min Cooling rate: 25°C/min	Sample sintered at T_1:1350°C, T_2:1300°C for 18 h attained the highest relative density (99.1%)	M_s: 27.4 emu/g for sample sintered at T_1 = 1350°C, T_2:1300°C for 18 h	[27]
NiCuZn ferrite	Commercial purchased NiCuZn ferrite powder and sintering	Commercial purchased NiCuZn ferrite	Sintering temperature: 850, 900, 950, 1000 and 1050°C Sintering time: 2 h	Density increased from 4.0 to 4.3 g/cm^3 Average grain sizes increased from 3.0 to 7.5 μm with increasing sintering temperatures	Initial permeability, μ_i, at 5 MHz increased with increased sintering temperature with maximum value of ~112 for sample sintered at 1050°C	[3]
$CoFe_2O_4$	Citrate precursor method	$Co(NO_3)_2.6H_2O$ $Fe(NO_3)_2.9H_2O$ $C_6H_8O_7$	Sintering temperature: 900, 1000 and 1100°C	Average grain sizes increased from 90.0 to 100.0 nm with increasing sintering temperatures	M_s increased from 53.7 emu/g to 74.5 emu/g	[4]

Material	Synthesis method	Starting materials	Sintering conditions	Microstructure features	Optimum magnetic properties	Reference
$Zn_{0.35}Ni_{0.57}Co_{0.03}Fe_{2.05}O_4$	Chemical combustion route	Metal nitrates Citric acid	Sintering temperature: 1050 and 1150°C Sintering time: 1 h	Average grain sizes increased from 0.61 to 0.94 μm with increasing sintering temperatures	Real part of the initial permeability increases with increasing sintering temperature from 85.2 to 209.7 at 10 kHz and from 90.4 to 238.1 at 1 MHz	[28]
$BaFe_{12}O_{19}$	High-energy ball milling	$BaCO_3$ Fe_2O_3	Sintering atmosphere: Air atmosphere Sintering temperature: 800, 900 and 1150°C Sintering time: 1 h	The highest measured density is 4.88 g/cm³ for the sample mechanically alloyed for 3 h and sintered at 1150°C	The highest Ms value of 63.57 emu/g was measured for the sample mechanically alloyed for 3 h and sintered at 1150°C The highest coercivity, H_C, value is 5.31 kOe obtained for the sample milled for 9 h and sintered at 900°C	[29]
$Ni_{0.3}Zn_{0.7}Fe_2O_4$	Sol–gel auto-combustion method	$Fe(NO_3)_3.9H_2O$ $Zn(NO_3)_2.\ 4H_2O$ $Ni(NO_3)_2.6H_2O$	Sintering atmosphere: Air atmosphere Sintering temperature: 350, 400, 500, 600, 800, 1000 and 1200°C Sintering time: 1 h	Average crystallite sizes increased from 13.0 to 58 nm with increasing sintering temperatures	Highest M_s of 49.4 emu/g for sample sintered at 1200°C with the largest crystallite size Maximum H_c of13.82 Oe was observed in sample sintered at 800°C (crystallite size:36 nm) due to transition of the magnetic single-domain to magnetic multi-domain structure	[24]

Material	Synthesis method	Starting materials	Sintering conditions	Microstructure features	Optimum magnetic properties	Reference
$Y_3Fe_5O_{12}$	Low temperature solid state reaction	$Y(NO_3)_3{\cdot}6H_2O$ $Fe(NO_3)_3{\cdot}9H_2O$ Citric acid	Sintering atmosphere: Air atmosphere Sintering temperatures: 1220, 1240, 1280 and 1320°C Sintering times: 3 h Heating rate:10°C/min	Optimum sintering temperature of 1280°C with highest value of ~5.08 g/cm³ that was about 98% of the XRD density (5.17 g/cm³)	M_s: 13.8 mT for sample sintered at 1280°C	[30]
$Mn_{0.49}Zn_{0.48}Fe_{2.06}O_4$	Co-precipitation method		Sintering atmosphere: air, mixture of nitrogen and air, and nitrogen atmospheres Sintering temperature: 850, 880, 900 and 950°C Heating rate: 5, 6, 7 and 8°C/min	Sintering in nitrogen produced pure $Mn_{0.49}Zn_{0.48}Fe_{2.06}O_4$ ferrite while sintered in air or mixture of air and nitrogen contained oxides such as Fe_2O_3, Mn_2O_3 and ZnO Highest sintering density of 4.82 g/cm³ and homogeneous grain size were found in sample sintered at 880°C in nitrogen atmosphere with 5°C/min heating rate	Largest M_s value of 90.02 emu/g was observed in sample sintered at 880°C in nitrogen atmosphere with 5°C/min heating rate The M_s decreased with increasing heating rate from 90.02 to 80.60 emu/g	[22]
$Ni_{0.35}Cu_{0.05}Zn_{0.60}Fe_2O_4$ $Mg_{0.35}Cu_{0.05}Zn_{0.60}Fe_2O_4$	Conventional mixed oxide method and microwave sintering	NiO MgO CuO ZnO Fe_2O_3	Sintering temperature: 850, 875, 900, 925, 950, 975 and 1000°C Sintering time: 30 min	Largest grains and highest density of 5.28 and 4.95 g/cm³ were observed in $Ni_{0.35}Cu_{0.05}Zn_{0.60}Fe_2O_4$ sintered at 950°C and $Mg_{0.35}Cu_{0.05}Zn_{0.60}Fe_2O_4$ sintered at 900°C, respectively	μ_i at 10 kHz showed maximum values of ~2825 for both $Ni_{0.35}Cu_{0.05}Zn_{0.60}Fe_2O_4$ sintered at 950°C and $Mg_{0.35}Cu_{0.05}Zn_{0.60}Fe_2O_4$ sintered at 900°C, respectively Highest M_s of ~129 and 88 emu/g for $Ni_{0.35}Cu_{0.05}Zn_{0.60}Fe_2O_4$ sintered at 950°C and $Mg_{0.35}Cu_{0.05}Zn_{0.60}Fe_2O_4$ sintered at 900°C, respectively	[31]

Material	Synthesis method	Starting materials	Sintering conditions	Microstructure features	Optimum magnetic properties	Reference
$Y_3Fe_5O_{12}$	Solid-state reaction method and microwave sintering	Fe_2O_3 Y_2O_3	**Conventional sintering (CS)** Sintering temperature: 1300°C Sintering time: 6 h Heating rate:2°C/min Cooling rate: 2°C/min **Microwave sintering (MS)** Sintering temperature: 900°C and 1000°C Sintering time: 20 min (for 900°C) and 30 min (for 1000°C) Heating rate: 8°C/min Cooling rate: 30°C/min	**Conventional sintering (CS)** Grain size: 3–5 μm Density: 98% T.D **Microwave sintering (MS)** Grain size: 1.5 μm (900°C), 5–10 μm (1000°C) Density: 96% T.D (900°C), 98% T.D (1000°C)	**Conventional sintering (CS)** M_s: 25.42 emu/g Coercive force, H_c: 25.36 Oe **Microwave sintering (MS)** M_s: 14.60 emu/g H_c: 34.82 Oe	[32]
$Mn_{1-x}Zn_xFe_2O_4$ (x = 0, 0.1, 0.2, 0.4)	Sol-gel combustion method	Fe_2O_3 SrO	Sintering atmosphere: Air atmosphere Sintering temperature: 700, 800 and 950°C Sintering time: 5 h	Average grain size: ~1–2 μm Density increased with increasing sintering temperatures from 4.64 to 4.80 g/cm^3	μ_i increased from 2100 to 2450 with increasing sintering temperatures. M_s: no sign of M_s even at 10000 Oe magnetic field	[33]

Table 1. Summary of synthesis methods, sintering conditions, microstructure features and optimum magnetic properties of ferrites.

crystallite size is observed to be significantly increased with prolong sintering time up to 7 h [23], thus enhancing the saturation magnetization of the material. This may be as a result of the improved crystallinity, which implying a better exchange interaction.

However, the effects of sintering conditions, particularly sintering temperatures and times, on magnetic properties evolution are not necessarily increased with increase in sintering temperature or sintering time. This is attributed to the resulting microstructure features such as abnormal grains and pores which are related to the decrease of density, thus decreasing the magnetic properties, mainly the volume magnetization and the magnetic induction. The non-linear relationship is also due to the characteristic of the transition from single-domain to multi-domain grains. This phenomenon is largely observed in coercivity value against particle or grain size of the magnetic material [24].

6. Results from experimental works on single-sample sintering (SSS) and multi-sample sintering (MSS) of NiZn Ferrites and YIG

6.1. Comparative study of single-sample and multi-sample sintering of NiZn ferrites

Sintering temperatures increments from 600 to 1400°C increase the average grain size in both MSS and SSS as has been shown in **Table 2**, resulting from several processes. Those processes involve particles rearrangement and formation of dumbbell-liked structure between the particles contact points or known as the necking process. The grains are formed when the particles move closer during intermediate sintering stage as the sintering temperature goes higher. Finally, pores near or on the grain boundaries are gradually removed through the diffusion of vacancies associated by the pores along the grain boundaries, having only slight densification of the sample. The average grain size between the two different schemes shows small but significant difference. The striking difference in the microstructure is seen in $Ni_{0.3}Zn_{0.7}Fe_2O_4$ sintered at 1100°C as shown in **Figure 3(a)** and **(b)**. The striking difference arises from the

Sintering temperature (°C)		600	700	800	900	1000	1100	1200	1300	1400
Average grain size (μm)	MSS	0.19	0.21	0.23	0.24	0.43	1.07	1.23	2.65	4.98
	SSS	0.13	0.15	0.19	0.23	0.30	0.39	1.05	2.08	5.35
Experimental density (g/cm³)	MSS	3.48	3.62	3.7	3.98	4.02	4.2	4.64	4.73	4.56
	SSS	4.23	4.49	4.62	4.73	4.78	4.81	4.91	4.93	4.88
Saturation induction, B_s (Gauss)	MSS	23.9	29.7	97.8	503.0	522.8	865.3	908.3	949.7	1076.0
	SSS	23.4	29.5	68.3	424.2	523.0	572.0	605.3	774.7	930.7
Coercivity, H_c (Oe)	MSS	3.4	7.7	12.5	5.0	4.1	1.3	1.2	0.5	0.4
	SSS	3.0	9.7	11.5	6.7	3.8	3.5	1.6	0.9	0.3

Table 2. Average grain size, experimental density, saturation induction and coercivity of $Ni_{0.3}Zn_{0.7}Fe_2O_4$ for different sintering temperatures.

http://dx.doi.org/10.5772/intechopen.68500

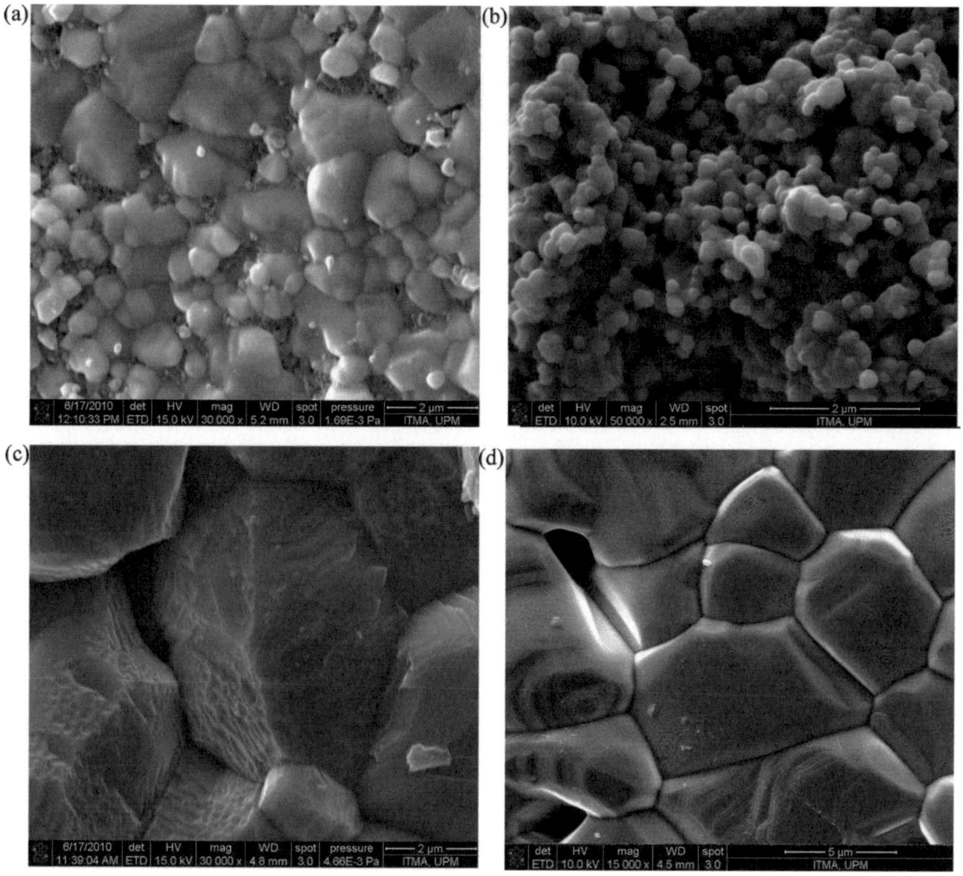

Figure 3. FESEM micrographs for $Ni_{0.3}Zn_{0.7}Fe_2O_4$ sintered at: (a) 1100°C (MSS), (b) 1100°C (SSS), (c) 1400°C (MSS) and (d) 1400°C (SSS).

different surface reactivities prior to the 1100°C sintering: for the MSS, the surface reactivity is high because the green bodies $Ni_{0.3}Zn_{0.7}Fe_2O_4$ are compacted from originally high-reactivity as-milled powders. However, the SSS is subjected to several times of repeated sintering, thus reducing the surface reactivity of the material prior to the 1100°C sintering. The pores which exist in MSS $Ni_{0.3}Zn_{0.7}Fe_2O_4$ sintered at 1400°C as shown in **Figure 3(c)** causing the decrease in density. Intragranular pores are trapped pores in the grains due to rapid grain growth and also probably due to zinc loss. The pores are known to be bad inclusions because they would pin down the domain wall, thus reducing the magnetization. However, no significant pores are observed in SSS $Ni_{0.3}Zn_{0.7}Fe_2O_4$ sintered at 1400°C. This is due to repeated sintering which could provide more time for the trapped pores to be removed; consequently, no significant amount of intragranular porosity was observed.

The focal question of what factors that subjected to different hysteresis shapes characteristic in both sintering treatments is of great interest. The shapes of the hysteresis loop are largely correlated with the microstructural features of the material, particularly the grains in the sample [34–40]. Besides, the disparities of the shapes also arise from the various grain shapes, grain sizes, compositions, strains, and imperfections present in the sample. Maximum magnetic induction, B_s, of nickel zinc ferrite could range from 1000 to 3000 G [20, 41]. The experimental values of B_s are shown in **Table 2** range from 23.9 to 1076.0 G for MSS and from 23.4 to 930.7 G for SSS. The various ranges of B_s are subjected to the influence of several reasons in which categorizing the B-H hysteresis loops into several groups. The noticeably different B-H hysteresis loops are seen as three different shapes in both MSS and SSS. The loops are divided into three groups based on their magnetic behavior: strongly, moderately and weakly ferromagnetic, which are known to be strongly influenced by microstructural properties, domain states, and crystallinity of the samples. $Ni_{0.3}Zn_{0.7}Fe_2O_4$ sintered from 600 to 800°C for both MSS and SSS as shown in **Figure 4(a)** and **(b)**, respectively, is classified as the first group with weakly ferromagnetic behavior. The shape of the hysteresis loops is affected by mixture phases of ferromagnetic and paramagnetic phase and also most likely by some superparamagnetic phase [42]. The significant amount of amorphous grain boundary

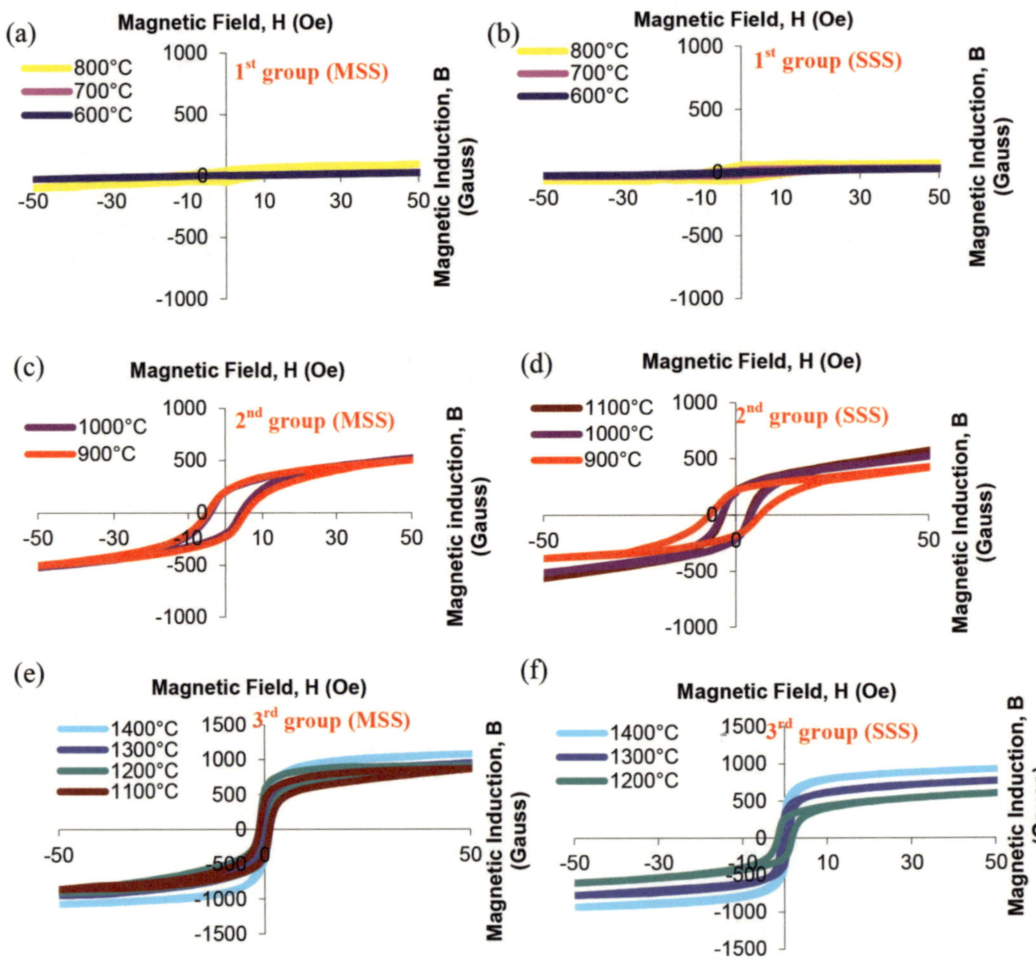

Figure 4. B-H hysteresis loops of $Ni_{0.3}Zn_{0.7}Fe_2O_4$ for first group of (a) MSS (b) SSS, second group of (c) MSS, (d) SSS, and third group of (e) MSS (f) SSS.

volumes has contributed to the paramagnetic phase, which arises from the fine grain size of the samples [36, 39, 42]. In addition, a superparamagnetic phase is contributed by the nano-sized grains. The shapes show a little hysteresis with narrowly bulging but linear-looking loops and have a very low Bs, indicating a very small amount of ferromagnetic phase. Due to the lower sintering temperature than the other two groups, the crystalline-phase percentage is small, while the amorphous-phase percentage is still significant. The grouping for the moderately ferromagnetic second group is slightly different between MSS and SSS where $Ni_{0.3}Zn_{0.7}Fe_2O_4$ sintered from 900 to 1000°C is for the MSS and from 900 to 1100°C is for SSS (see **Figures 4(c)** and **(d)**). The difference between the two sintering schemes is due to the influence of microstructural properties (see **Figure 3**). $Ni_{0.3}Zn_{0.7}Fe_2O_4$, which belongs to this group, shows a slanted sigmoid shape which is recognized to demonstrate moderate ferromagnetic behavior with negligible paramagnetic behavior since there is still remained a significant amount of the amorphous phase. The B-H loops of the MSS for this group have significantly higher B_s (M_s) values but falling Hc values (see **Figure 5**) indicating, respectively, higher ferromagnetic phase crystallinity and starting dominance of multi-domain magnetization-demagnetization processes. However, in the SSS, $Ni_{0.3}Zn_{0.7}Fe_2O_4$ still exhibits single domain grains as shown in **Figure 5**. Consequently, the magnetization of the SSS samples is largely exhibiting via spin rotation, thus lowering the magnetization values than that of MSS samples which already possessing multi-domain grains though sintered at similar sintering temperature. The result is clearly observed in $Ni_{0.3}Zn_{0.7}Fe_2O_4$ sintered at 1100°C: while the MSS sample already behaving as strongly ferromagnetic (third group), the SSS sample is still belong to the second group. The third group displays strongly ferromagnetic behavior (B_s, M_s) with a diminishing amorphous phase due to insignificant amount of amorphous grain boundaries volume. This behavior is exhibited by $Ni_{0.3}Zn_{0.7}Fe_2O_4$ sintered at much higher sintering temperature. The sintering temperatures range from 1100 to 1400°C for the MSS and from 1200 to 1400°C for the SSS. The well-known erect, narrower and well-defined sigmoid shape has been observed for the third group of hysteresis loops. This strongly ferromagnetic behavior is contributed by very high crystallinity, high density with a minute amount of microstructural defects, and large size of grains, resulting from high sintering temperature. Therefore, the combinations of these particular parameters would allow domain walls movement to become easier in the magnetization and demagnetization process.

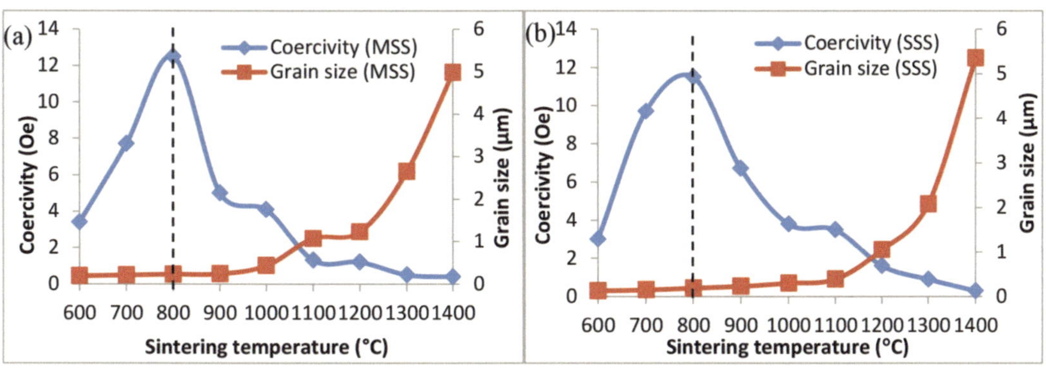

Figure 5. Coercivity and grain size as a function of sintering temperatures for (a) MSS and (b) SSS of $Ni_{0.3}Zn_{0.7}Fe_2O_4$ samples.

The H_c values in **Figures 5(a)** and **(b)** are found to increase as the sintering temperature increased from 600 to 800°C, reaches a maximum value and decreased from 800 to 1400°C. Interestingly, the similar trend is showed in both MSS and SSS schemes, proving the transition of single-domain to multi-domain grains happened in the similar grain size range. The drop of H_c values in SSS has occurred earlier for $Ni_{0.3}Zn_{0.7}Fe_2O_4$ having grain size of 0.19 μm as compared to MSS where it drops at 0.23 μm. H_c is probably the property most sensitive to porosity and grain size [20] nevertheless to the anisotropy field as well. Soft ferrites with nanometric grains exhibit a much higher H_c than samples having grain sizes of the order of few microns. An inversely proportional trend of H_c against grain size is observed for multi-domain grains, which consisting of more domain walls. Therefore, the contribution of lower energy domain walls movement to demagnetization or demagnetization than that of domain rotation increases. Consequently, coarse grains are expected to display low H_c [43]. However, below a certain size, which the H_c reaches a maximum value, or known as the critical size, the grains are single-domain grains [44, 45]. The increasing values of coercivity for lower sintering (≤800°C) were due to size-shape anisotropy (necking phase in the microstructure) and magnetocrystalline anisotropy. For higher sintering temperatures (≥900°C), the grain size exceeded the critical grain size with the disappearing size-shape anisotropy but with remaining magnetocrystalline anisotropy. Magnetocrystalline anisotropy is reduced in larger grains by decreasing the internal stress and crystal anisotropy [46], helping in better domain walls movement, thus decreasing the H_c. Within this grain size range, the anisotropy and defects including pores govern the H_c values. **Figure 5(a)** and **(b)** greatly affirms the trend, giving a maximum H_c of 12.5 Oe at 0.23 μm and 11.5 Oe 0.19 μm for MSS and SSS, respectively. Therefore, the range of critical size for $Ni_{0.3}Zn_{0.7}Fe_2O_4$ is approximately from 0.20 to 0.25 μm.

Figures 6 and **7** present the real part of permeability with frequency dispersion from 1 MHz to 1.8 GHz for both sintering schemes. Generally, the permeability is related to two different magnetizing mechanisms which are spin rotational and domain wall movement. Normally, spin rotation occurs at higher frequency when domain is damped and could not follow the applied electromagnetic wave. According to Snoek's law [47], the relation between resonance frequency f_R and the initial permeability μ_i for Ni–Zn ferrites may be expressed as follows:

$$f_R = (1/\mu_i) \times 3 \times 10^9 \text{ Hz} \tag{1}$$

This indicates that the lower the initial permeability values, the higher will be the frequency at which resonance phenomenon occurs. The value of real part of permeability for $Ni_{0.3}Zn_{0.7}Fe_2O_4$ sintered at 600 and 700°C for both sintering schemes is independent of frequency in this measured frequency region (1 MHz–1.8 GHz), whereas $Ni_{0.3}Zn_{0.7}Fe_2O_4$ sintered at 800°C only showed the dependency at about 100 MHz before reach the onset of resonance frequency. The resonance frequency is mostly observable in samples with lower sintering temperatures with the presence of single phase $Ni_{0.3}Zn_{0.7}Fe_2O_4$ (800–1000°C) for both sintering schemes. The coarsened grains for $Ni_{0.3}Zn_{0.7}Fe_2O_4$ sintered at 1100°C and above would lead to a ferromagnetic resonance at a lower frequency, in which, therefore, the resonance frequency could not be observed in the permeability spectra within the frequency region. At high frequencies, the domain walls cannot keep pace with the rapidly changing magnetic field, decreasing the value of real part of permeability. In powdered ferrites where each grain contains only a few domains, magnetization process occurs primarily by domain rotation and less by domain wall movement [48].

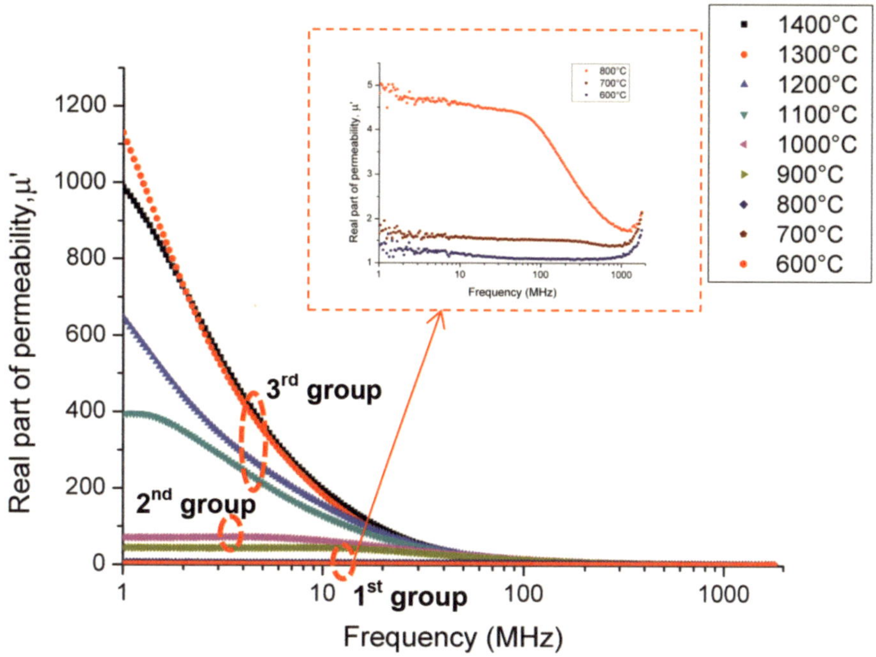

Figure 6. Graph of real permeability, μ′ against frequency for $Ni_{0.3}Zn_{0.7}Fe_2O_4$ multi-sample sintering.

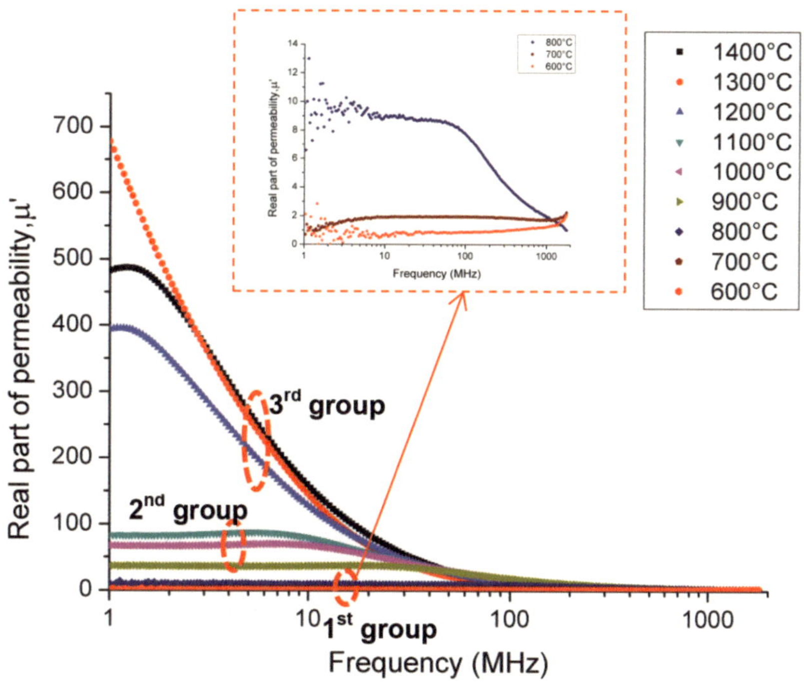

Figure 7. Graph of real permeability, μ′against frequency for $Ni_{0.3}Zn_{0.7}Fe_2O_4$ single-sample sintering.

The frequency stability for real permeability is varied from one group to another, which therefore varying the suitable applications for each group. The resonance frequency represents the high-frequency limit up to which the material can be used in a device. $Ni_{0.3}Zn_{0.7}Fe_2O_4$ with strong ferromagnetic behavior is suitable for lower frequency application (less than 1 MHz) because of the frequency stability at lower frequency where resonance frequency is found to be lowered than measured frequency range. Some applications that are operating in the frequency range of 0.5–5 MHz are ferrite antennas for medium and long wave broadcast bands, power transformers, and cores for electromagnetic suppression. For moderate ferromagnetic behavior, $Ni_{0.3}Zn_{0.7}Fe_2O_4$ sintered through MSS displays ferromagnetic resonance at frequency of 10.7 MHz (sintered at 900°C) and 3.9 MHz (sintered at 1000°C) with maximum real part of permeability value of 44.4 and 72.1, respectively, whereas for $Ni_{0.3}Zn_{0.7}Fe_2O_4$ sintered through SSS, samples demonstrate ferromagnetic resonance at frequency of 20.1 MHz (sintered at 900°C), 6.31 MHz (sintered at 1000°C), 5.40 MHz (sintered at 1100°C) with maximum real part of permeability value of 37.4, 70.5 and 87.9, respectively. The materials could be used for the application of the solid core of inductors for resonant circuits or transformers operating in the approximate frequency range 2–20 MHz [41, 49], ferrite antennas for short wave broadcast bands, power transformers for the approximate frequency range 2–30 MHz and cores for electromagnetic interference suppression [41]. For weak ferromagnetic behavior, only $Ni_{0.3}Zn_{0.7}Fe_2O_4$ sintered at 800°C for both sintering schemes displays the resonance phenomenon. The frequencies are stabled until 44.3 and 39.5 MHz for SSS and MSS, respectively, with maximum real part of permeability value of 8.64 and 4.42, respectively. Permeability with a value less than 12 is used for inductors and for resonant circuits operating at frequencies above 30 MHz and cores for electromagnetic interference suppression, whereas a much higher frequency than 1.8 GHz is needed to show resonance behavior in $Ni_{0.3}Zn_{0.7}Fe_2O_4$ sintered at 600 and 700°C due to smaller grain size and lower magnetic mass in the $Ni_{0.3}Zn_{0.7}Fe_2O_4$.

The complex permeability could also be classified into three different groups which are value-differentiated groups: strongly, moderately and weakly ferromagnetic behavior. By determining the critical size of single-domain to multi-domain grains through plot in **Figure 5**, it is found that $Ni_{0.3}Zn_{0.7}Fe_2O_4$ with weakly ferromagnetic behavior contains less than 50% multi-domain grains ($Ni_{0.3}Zn_{0.7}Fe_2O_4$ sintered from 600 to 800°C for MSS and SSS), whereas $Ni_{0.3}Zn_{0.7}Fe_2O_4$ with the moderately ferromagnetic behavior possesses more than 50% multi-domain grains ($Ni_{0.3}Zn_{0.7}Fe_2O_4$ sintered from 900 to 1000°C for MSS; $Ni_{0.3}Zn_{0.7}Fe_2O_4$ sintered from 900 to 1100°C for SSS), and 100% of the grains are multi-domain grains ($Ni_{0.3}Zn_{0.7}Fe_2O_4$ sintered from 1100 to 1400°C for MSS; $Ni_{0.3}Zn_{0.7}Fe_2O_4$ sintered from 1200 to 1400°C for SSS) which are counted as strongly ferromagnetic behavior. Therefore, $Ni_{0.3}Zn_{0.7}Fe_2O_4$ sintered below 800°C (with grain size less than 0.25 μm) is dominated by spin rotation, whereas $Ni_{0.3}Zn_{0.7}Fe_2O_4$ sintered from 800°C upwards dominated by domain wall movement and spin rotation. The reason for the increase in permeability with sintering temperature is attributed to the increase of grain size and reduction of porosity, reducing the anisotropy arising from the demagnetizing fields outside of grains. Fewer number of the grain boundaries would be present in $Ni_{0.3}Zn_{0.7}Fe_2O_4$ sintered at high temperatures, causing the existence of very mobile domain walls thus increasing the permeability value. Moreover, during grain growth, many pores would be removed, thus reducing the hindrance to the domain walls motion because pores provide stress concentration that may affect the magnetization's easy direction. However, the decrease in the real part of the permeability for $Ni_{0.3}Zn_{0.7}Fe_2O_4$ sintered at 1400°C is attributed to zinc loss [47] and existence of pores (see **Figure 3**).

The loss factor is observed to increase with a rise of the frequency from 1 MHz and attain the maximum value at a particular frequency and decreased with a further increase in frequency. The loss factor values increase with increasing sintering temperature in both MSS and SSS as shown in **Figures 8** and **9** for MSS and SSS, respectively. The frequency at which losses begin to increase due to the onset of resonance varies with the sintering temperatures from 2 to 100 MHz for both sintering schemes. As the sintering temperatures increase, the domain walls movement becomes easier in the larger grain, thus inducing larger eddy current. It is caused by the changing magnetic fields inside the sample which give rise to circulating currents inside the sample, and hence to energy losses [50]. However, in $Ni_{0.3}Zn_{0.7}Fe_2O_4$, eddy current losses are not dominant due to its high electrical resistivity. The larger grain raises the number and size of magnetic domains which contribute to loss due to delay in domain wall motion. The losses in ferrites are associated with domain wall relaxation and rotational resonance. In higher frequency regions (>500 MHz), most of the domain walls are damped and become less important as spin rotational would continue to occur [20].

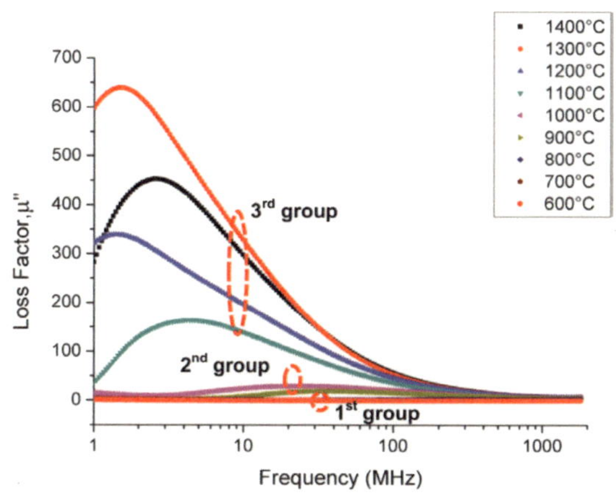

Figure 8. Graph of loss factor, μ" against frequency for $Ni_{0.3}Zn_{0.7}Fe_2O_4$ multi-sample sintering.

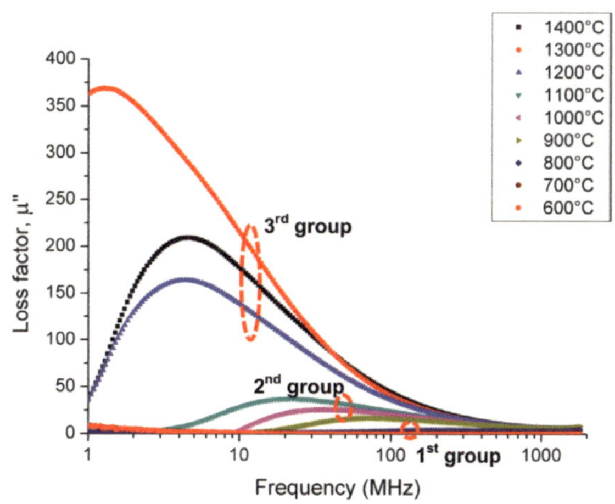

Figure 9. Graph of loss factor, μ" against frequency for $Ni_{0.3}Zn_{0.7}Fe_2O_4$ single-sample sintering.

6.2. Comparative study of single-sample and multi-sample sintering of yttrium iron garnet

A systematic track of microstructure-magnetic properties evolution of several polycrystalline Yttrium iron garnet (YIG) ferrite samples as a result of different sintering schemes was investigated in detail, focusing on the attendant occurrence of their dependency: an aspect seemingly ignored, hitherto in the garnet ferrite previous literatures for the past eight decades.

In order to prepare Yttrium iron garnet (YIG) ferrite sample, Fe_2O_3 (Alfa Aesar, 99.945%) and Y_2O_3 (Alfa Aesar, 99.99%) powders were weighed and mixed according to the stoichiometric proportions required in the final YIG samples based on the reaction:

$$3Y_2O_3 + 5Fe_2O_3 \rightarrow 2Y_3Fe_5O_{12} \quad (2)$$

The powder then mechanically alloyed into nanosize via mechanical alloying technique. Two batches of samples were produced with different sintering scheme: SSS and MSS, each covering a range of low sintering temperature from 600°C up to high sintering temperature of 1400°C with 100°C increments. The samples were analyzed by using a LEO 912AB energy filter transmission electron microscope (TEM), Philips Expert PW3040 diffractometer operating at 40 kV/30 mA using Cu Kα radiation, scanning electron microscopy (SEM), MATS-2010S Static Hysteresis Graph at room temperature under applied magnetic fields 0–50 Oe (0–4000 A/m) and HP4291B Materials Impedance Analyzer at room temperature for their evolution stage in crystalline phases, microstructure, magnetic hysteresis-loop parameters, and magnetic permeability components, respectively.

With great experimental care, both the SSS and MSS batches yielded similar variation of microstructure-magnetic properties evolution (**Table 3**). The results showed an increasing tendency of

Sintering temperature, (T, °C)	Single-sample sintering				Multi-sample sintering			
	Grain size, (±0.01 μm)	Saturation induction, B_s (Gauss)	Saturation magnetization, M_s (emu/cm³)	Coercivity, H_c (Oe)	Grain size, (±0.01 μm)	Saturation induction, B_s (Gauss)	Saturation magnetization, Ms (emu/cm³)	Coercivity, H_c (Oe)
600	0.16	2.1	1.7	2.6	0.20	16.9	2.2	0.1
700	0.17	2.3	2.6	3.1	0.21	24.9	2.7	0.7
800	0.18	16.4	3.9	6.7	0.25	35.8	4.7	1.3
900	0.28	20.6	4.8	10.9	0.26	49.6	5.2	3.9
1000	0.33	120.7	5.4	15.5	0.28	128.6	5.5	15.8
1100	0.60	173.2	5.7	18.5	0.58	185.7	6.2	19.3
1200	1.14	223.7	10.9	12.4	0.80	244.5	12.8	15.2
1300	1.68	378.9	21.2	7.4	1.25	463.1	23.3	8.8
1400	2.71	570.4	26.3	4.3	3.09	714.6	29.1	2.9

Table 3. Microstructural and magnetic parameters of single- and multi-sample sintering YIG samples with various sintering temperature variations.

the saturation magnetization and magnetic induction with grain size, which was attributed to increase of crystallinity and demagnetizing field reduction in the grains. The variation in coercivity corresponded to the changes of anisotropy field within the samples due to grain size changes. Specifically, the starting appearance of room temperature ferromagnetic order suggested by the sigmoid-shaped B-H loops seems to be dependent on a sufficient number of large enough magnetic-domain containing grains formed in the microstructure. Viewed simultaneously, the B-H loops (appeared to be belonging to three groups with different magnetism-type dominance, respectively dependent on phase purity and distribution of grain size. The clearly tracked evolution of the hysteresis (**Figures 10** and **11**) and permeability component (**Figure 12**) strongly suggests that high reactivity grain surfaces and great-care human handling of the sample preparation process contributed to the startlingly clear microstructure-property evolution trends.

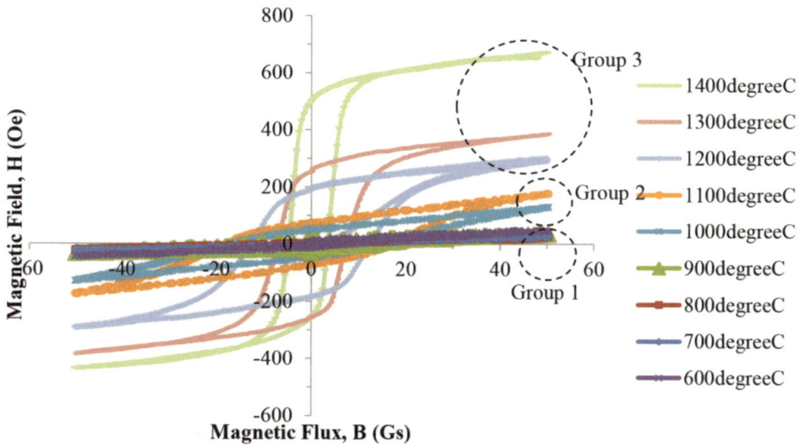

Figure 10. B-H hysteresis loop for single-samples sintered at various temperatures. The circles inside the figure indicate 3 different groups of B-H curve evolution.

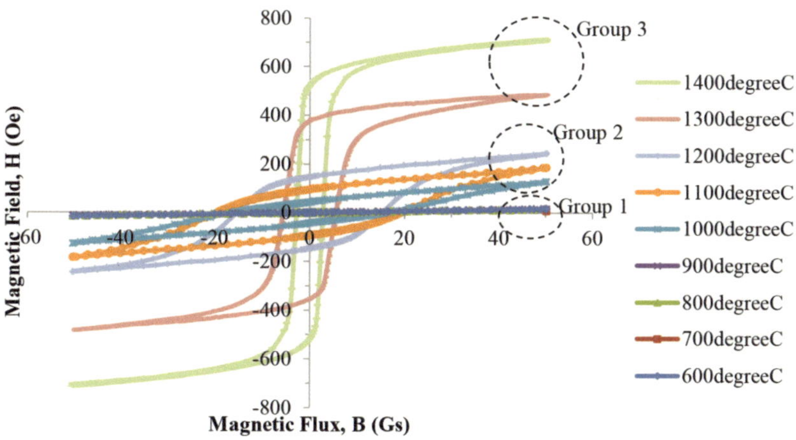

Figure 11. B-H hysteresis loop for multi-samples sintered at various temperatures. The circles inside the figure indicate 3 different groups of B-H curve evolution.

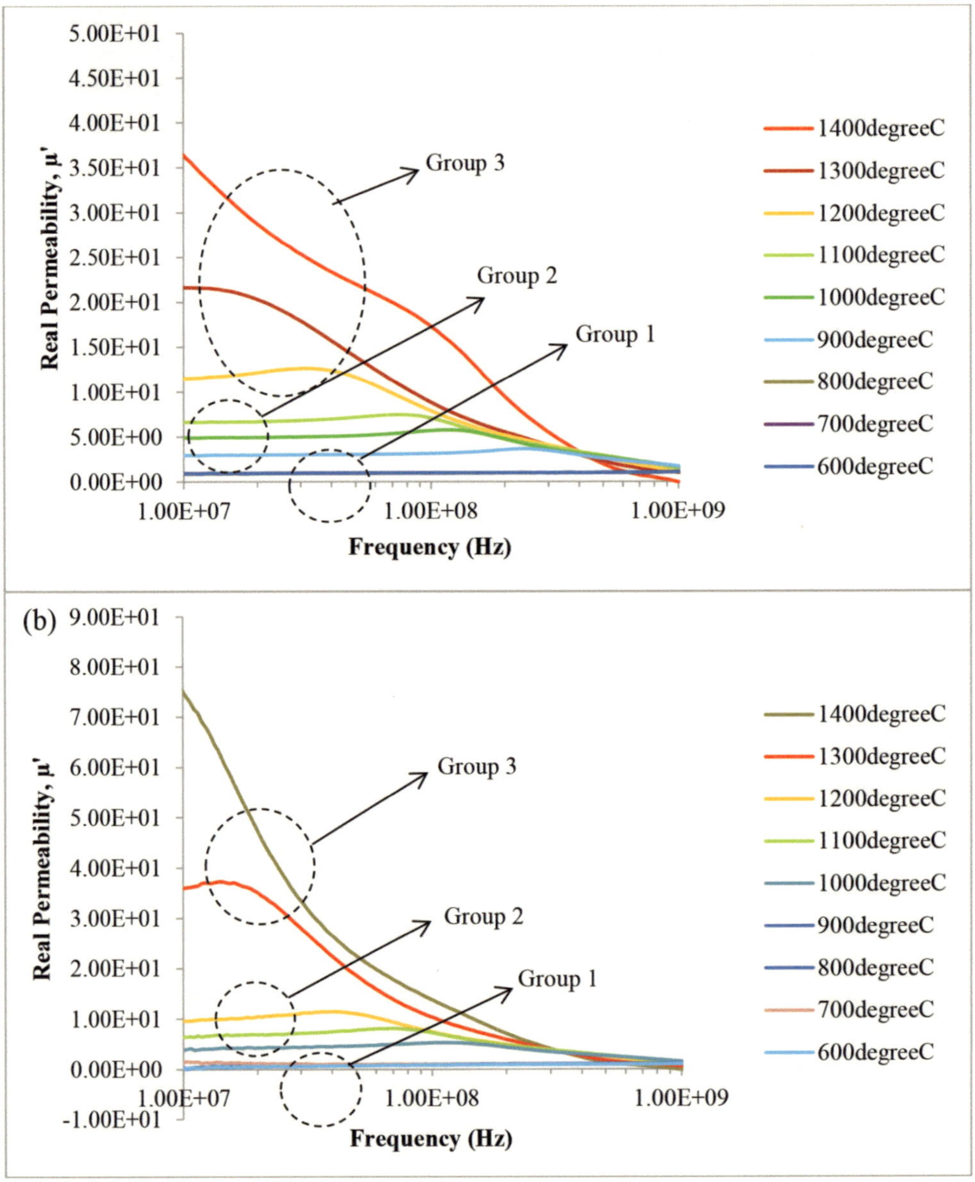

Figure 12. Real permeability measured at room temperature in range 10 MHz to 1 GHz (a) single-sample sintering; (b) multi-samples sintering samples.

Acknowledgements

We would like to dedicate this chapter and show our gratitude to the late Assoc. Prof. Dr. Mansor Hashim from Universiti Putra Malaysia, Malaysia for sharing his pearls of wisdom with us during the course of this research.

Author details

Ismayadi Ismail[1]*, Idza Riati Ibrahim[1] and Rodziah Nazlan[2]

*Address all correspondence to: ismayadi@upm.edu.my

1 Materials Synthesis and Characterisation Laboratory, Institute of Advanced Technology, Universiti Putra Malaysia, Serdang, Selangor, Malaysia

2 Department of Material Technology, Faculty of Industrial Science and Technology, Universiti Malaysia Pahang, Kuantan, Pahang, Malaysia

References

[1] Cao G, Wang Y. Nanostructures and Nanomaterials: Synthesis, Properties, and Applications. 2nd ed. Toh Tuck Lin: World Scientific. 2011

[2] Winterer M. Nanocrystalline Ceramics Synthesis and Structure. New York: Springer-Verlag Berlin Heidelberg. 2002

[3] Yan YI, Ngo KDT, Hou D, Mu M, Mei Y, Lu G. Effect of sintering temperature on magnetic core-loss properties of a NiCuZn ferrite for high-frequency power converters. Journal of Electronic Materials. 2015;**44**(10):3788-3794. DOI: 10.1007/s11664-015-3836-z

[4] Rana K, Thakur P, Sharma P, Tomar M, Gupta V, Thakur A. Improved structural and magnetic properties of cobalt nanoferrites : Influence of sintering temperature. Ceramics International. 2015;**41**:4492-4497. DOI: http://dx.doi.org/10.1016/j.ceramint.2014.11.143

[5] Huang CC, Hung YH, Huang JY, Kuo MF. Impact of stoichiometry and sintering temperature on magnetic properties of $Y_3Mn_xAl_{0.83-x}Fe_{4.17}O_{12}$ ferrites. IEEE Transactions on Magnetics. 2014;**50**(8):5-8. DOI: 10.1109/TMAG.2014.2310703

[6] Sujatha C, Reddy KV, Babu KS, Reddy AR, Rao KH. Effect of sintering temperature on electromagnetic properties of NiCuZn ferrite. Ceramics International. 2013;**39**(3):3077-3086. DOI: http://dx.doi.org/10.1016/j.ceramint.2012.09.087

[7] Islam R, Rahman O, Hakim MA, Saha DK. Effect of Sintering. Temperature on structural and magnetic properties of $Ni_{0.55}Zn_{0.45}Fe_2O_4$ ferrites. Materials Sciences and Applications. 2012;**3**:326-331. DOI: http://dx.doi.org/10.4236/msa.2012.35048

[8] Shirsath SE, Kadam RH, Gaikwad AS, Ghasemi A, Morisako A. Effect of sintering temperature and the particle size on the structural and magnetic properties of nanocrystalline $Li_{0.5}Fe_{2.5}O_4$. Journal of Magnetism and Magnetic Materials. 2011;**323**:3104-3108. DOI: http://dx.doi.org/10.1016/j.jmmm.2011.06.065

[9] Liu C, Lan Z, Jiang X, Yu Z, Sun K, Li L, et al. Effects of sintering temperature and Bi_2O_3 content on microstructure and magnetic properties of LiZn ferrites. Journal of Magnetism and Magnetic Materials. 2008;**320**:1335-1339. DOI: http://dx.doi.org/10.1016/j.jmmm.2007.10.016

[10] Kingery WD, Bowen HK, Uhlmann DR. Introduction to Ceramics. 2nd ed. New York: John Wiley & Sons Limited. 1976

[11] Tan W, Cuellar C, Margueron X, Idir N. A common-mode choke using Toroid-EQ mixed structure. IEEE Transactions on Power Electronics. 2013;**28**(1):31-35. DOI: 10.1109/TPEL.2012.2205708

[12] Waqas H, Hussain A, Muhammad Q. Effect of firing temperature on the electromagnetic properties of electronic transformer cores developed by using Nanosized Mn–Zn Ferrite powders. Acta Metallurgica Sinica (English Letters). 2015;**28**(2):159-163. DOI: 10.1007/s40195-014-0180-x

[13] Mcdonald T. Design and characterization of a miniature varactor tuned ferrite rod antenna for 4 to 24 MHz. In: 2013 Loughborough Antennas & Propagation Conference. IEEE. 2013. pp. 250-254

[14] Dar MA, Majid K, Hanief M, Kotnala RK, Shah J. Synthesis and characterization of Li0.5Fe2.5-xGdxO4 ferrite nano-particles as a potential candidate for microwave device applications. Materials and Design. 2016;**90**:443-452. DOI: http://dx.doi.org/10.1016/j.matdes.2015.10.151

[15] Ramesh T, Shinde RS, Murthy SR. Synthesis and characterization of nanocrystalline $Ni_{0.94}Co_{0.03}Mn_{0.04}Cu_{0.03}Fe_{1.96-x}Al_xO_4$ ferrites for microwave device applications. Journal of Magnetism and Magnetic Materials. 2013;**345**:276-281. DOI: http://dx.doi.org/10.1016/j.jmmm.2013.06.041

[16] Cheng YJ, Member S, Huang QD, Wang YR, Li JL. Narrowband substrate integrated waveguide isolators. IEEE Microwave and Wireless Component Letters 2014;**24**(10): 698-700. DOI: 10.1109/LMWC.2014.2344440

[17] Geiler AL, Harris VG. Atom magnetism: Ferrite circulators—past, present, and future. IEEE Microwave Magazine. 2014;**15**(6):66-72. DOI: 10.1109/MMM.2014.2332411

[18] Yang X, Wu J, Gao Y, Nan T, Zhou Z, Beguhn S, Liu M, Sun NX. Compact and low loss phase shifter with low bias field using partially magnetized ferrite. IEEE Transactions on Magnetics. 2013;**49**(7):3882-3885. DOI: 10.1109/TMAG.2013.2244860

[19] Goldman A. Handbook of Modern Ferromagnetic Materials. Massachusetts: Kluwer Academic Publishers. 1999

[20] Goldman A. Modern Ferrite Technology. Pittsburgh: Springer Science and Business Media, Inc. 2006

[21] Gholizadeh A, Jafari E. Effects of sintering atmosphere and temperature on structural and magnetic properties of Ni-Cu-Zn ferrite nano-particles: Magnetic enhancement by a reducing atmosphere. Journal of Magnetism and Magnetic Materials. 2017;**422**:328-336. DOI: http://dx.doi.org/10.1016/j.jmmm.2016.09.029

[22] Wang GX, Wu HX, Jiang YF, Peng HF, Wang X. Effect of sintering conditions on the properties of Mn-Zn Ferrites produced by Co-precipitation. Advanced Materials Research. 2011;**391-392**:839-843. DOI: 10.4028/www.scientific.net/AMR.391-392.839

[23] Sun K, Liu H, Yang Y, Yu Z, Chen C, Wu G, Jiang X, Lan Z, Li L. Contribution of magnetization mechanisms in nickel-zinc ferrites with different grain sizes and its temperature relationship. Materials Chemistry and Physics. 2016;**175**:131-137. DOI: http://dx.doi.org/10.1016/j.matchemphys.2016.03.002

[24] Rahimi M, Kameli P, Ranjbar M, Salamati H. The effect of sintering temperature on evolution of structural and magnetic properties of nanostructured $Ni_{0.3}Zn_{0.7}Fe_2O_4$ ferrite. Journal of Nanoparticle Reseacrh. 2013;**15**:1865-1876. DOI: 10.1007/s11051-013-1865-1

[25] Sun L, Zhang R, Wang Z, Ju L, Cao E, Zhang Y. Structural, dielectric and magnetic properties of $NiFe_2O_4$ prepared via sol-gel auto-combustion method. Journal of Magnetism and Magnetic Materials. 2017;**421**:65-70. DOI: http://dx.doi.org/10.1016/j.jmmm.2016.08.003

[26] Thakur A, Kumar P, Thakur P, Rana K, Chevalier A, Mattei JL, et al.: Enhancement of magnetic properties of $Ni0.5Zn0.5Fe_2O_4$ nanoparticles prepared by the co-precipitation method. Ceramics International. 2016;**42**(9):10664-10670. DOI: http://dx.doi.org/10.1016/j.ceramint.2016.03.173

[27] Li XX, Zhou JJ, Deng JX, Zheng H, Zheng L, Zheng P, Qin HB. Synthesis of dense, fine-grained YIG ceramics by two-step sintering. Journal of Electronic Materials. 2016;**45**(10):4973-4978. DOI: 10.1007/s11664-016-4690-3

[28] Ghodake JS, Shinde TJ, Patil RP, Patil SB, Suryavanshi SS: Initial permeability of Zn-Ni-Co ferrite. Journal of Magnetism and Magnetic Materials. 2015;**378**:436-439. DOI: http://dx.doi.org/10.1016/j.jmmm.2014.11.041

[29] Sadullahoglu G, Ertug B, Gökçe H, Altuncevahir B, Öztürk M, Topkaya R, Akdogan N, Öveçoglu ML, Addemir O. The effect of milling time and sintering temperature on crystallization of $BaFe_{12}O_{19}$ phase and magnetic properties of Ba-hexaferrite magnet. Acta Physica Polonica A. 2015;**128**(3):377-382. DOI: dx.doi.org/10.12693/APhysPolA.128.377

[30] Yu H, Zeng L, Lu C, Zhang W, Xu G. Synthesis of nanocrystalline yttrium iron garnet by low temperature solid state reaction. Materials Characterization. 2011;**62**(4):378-381. DOI: http://dx.doi.org/10.1016/j.matchar.2011.02.002

[31] Penchal Reddy M, Madhuri W, Venkata Ramana M, Ramamanohar Reddy N, Siva Kumar K V., Murthy VRK, Siva Kumar K, Ramakrishna Reddy R. Effect of sintering temperature on structural and magnetic properties of NiCuZn and MgCuZn ferrites. Journal of Magnetism and Magnetic Materials. 2010;**322**(19):2819-2823. DOI: http://dx.doi.org/10.1016/j.jmmm.2010.04.036

[32] Yang Q, Zhang H, Liu Y, Wen Q, Jia L. The magnetic and dielectric properties of microwave sintered yttrium iron garnet (YIG). Materials Letters. 2008;**62**(17-18):2647-2650. DOI: 10.1016/j.matlet.2008.01.040

[33] Azadmanjiri J: Preparation of Mn—Zn ferrite nanoparticles from chemical sol—gel combustion method and the magnetic properties after sintering. Journal of Non-Crystalline Solids. 2007;**353**:4170-4173. DOI: 10.1016/j.jnoncrysol.2007.06.046

[34] Nazlan R, Hashim M, Ibrahim IR, Idris FM, Wan Ab Rahman WN, Abdullah NH, Ismail I, Kanagesan S, Abbas Z, Azis RS. Influence of indium substitution and microstructure

changes on the magnetic properties evolution of $Y_3Fe_{5-x}In_xO_{12}$. Journal of Materials Science: Materials in Electronics. 2015;**26**(6):3596-3609. DOI: 10.1007/s10854-015-2874-x

[35] Low ZH, Hashim M, Ismail I, Kanagesan S, Shafie MSE, Mohd Idris F, Ibrahim IR. Development of magnetic B-H hysteresis loops through stages of microstructure evolution of bulk $BaFe_{12}O_{19}$. Journal of Superconductivity and Novel Magnetism. 2015;**28**:3075-3086. DOI: 10.1007/s10948-015-3099-1

[36] Shafie MSE, Hashim M, Ismail I, Kanagesan S, Fadzidah MI, Idza IR, Hajalilou A, Sabbaghizadeh R. Magnetic M–H loops family characteristics in the microstructure evolution of $BaFe_{12}O_{19}$. Journal of Materials Science: Materials in Electronics. 2014;**25**(9):3787-3794. DOI: 10.1007/s10854-014-2090-0

[37] Ibrahim IR, Hashim M, Nazlan R, Ismail I, Kanagesan S, Wan Ab Rahman WN, Abdullah NH, Mohd Idris F, Bahmanrokh G. A comparative study of different sintering routes effects on evolving microstructure and B–H magnetic hysteresis in mechanically-alloyed Ni–Zn ferrite, $Ni_{0.3}Zn_{0.7}Fe_2O_4$. Journal of Materials Science: Materials in Electronics. 2014;**26**(1):59-65. DOI: 10.1007/s10854-014-2362-8

[38] Ismail I, Hashim M, Kanagesan S, Ibrahim IR, Nazlan R, Wan Ab Rahman WN, Abdullah NH, Mohd Idris F, Bahmanrokh G, Shafie MSE, Manap M. Evolving microstructure, magnetic properties and phase transition in a mechanically alloyed $Ni_{0.5}Zn_{0.5}Fe_2O_4$ single sample. Journal of Magnetism and Magnetic Materials. 2014;**351**:16-24. DOI: dx.doi.org/10.1016/j.jmmm.2013.09.041

[39] Nazlan R, Hashim M, Ibrahim IR, Ismail I. Dependence of magnetic hysteresis on evolving Single-Sample sintering in fine-grained Yttrium iron garnet. Journal of Superconductivity and Novel Magnetism. 2013;**27**(2):631-639. DOI: 10.1007/s10948-013-2328-8

[40] Bean CP. Hysteresis loops of mixtures of ferromagnetic micropowders. Journal of Applied Physics. 1955;**26**:1381-1383. DOI: http://dx.doi.org/10.1063/1.1721912

[41] Snelling EC. Soft ferrites: Properties and Applications. Massachusetts: Butterworths Publishing. 1985

[42] Ibrahim IR, Hashim M, Nazlan R, Ismail I, Wan Ab Rahman WN. Influence of evolving microstructure on magnetic-hysteresis characteristics in polycrystalline nickel–zinc ferrite, $Ni_{0.3}Zn_{0.7}Fe_2O_4$. Materials Research Bulletin. 2012;**47**(6):1345-1352. DOI: 10.1016/j.materresbull.2012.03.007

[43] Jahanbin T, Hashim M, Matori KA, Waje SB. Influence of sintering temperature on the structural, magnetic and dielectric properties of $Ni_{0.8}Zn_{0.2}Fe_2O_4$ synthesized by co-precipitation route. Journal of Alloys and Compounds. 2010;**503**(1):111-117. DOI: http://dx.doi.org/10.1016/j.jallcom.2010.04.212

[44] Azizi A, Sadrnezhaad SK. Effects of annealing on phase evolution, microstructure and magnetic properties of mechanically synthesized nickel-ferrite. Ceramics International. 2010;**36**(7):2241-2245. DOI: http://dx.doi.org/10.1016/j.ceramint.2010.06.004

[45] George M, John AM, Nair, S. S., Joy PA, Anantharaman MR. Finite size effects on the structural and magnetic properties of sol–gel synthesized $NiFe_2O_4$ powders. Journal of Magnetism and Magnetic Materials. 2006;**302**(1):190-195. DOI: http://dx.doi.org/10.1016/j.jmmm.2005.08.029

[46] Verma A, Goel TC, Mendiratta RG. Frequency variation of initial permeability of NiZn ferrites prepared by the citrate precursor method. Journal of Magnetism and Magnetic Materials. 2000;**210**:274-278. DOI: http://dx.doi.org/10.1016/S0304-8853(99)00451-5

[47] Nakamura T. Low-temperature sintering of Ni–Zn–Cu ferrite and its permeability spectra. Journal of Magnetism and Magnetic Materials. 1997;**168**:285-291. DOI: http://dx.doi.org/10.1016/S0304-8853(96)00709-3

[48] Naughton BT, Majewski P, Clarke DR. Magnetic properties ofnickel–zinc ferrite toroids prepared from nanoparticles. Journal of American Ceramic Society. 2007;**90**:3547-3553. DOI: 10.1111/j.1551-2916.2007.01981.x

[49] Warne DF. Newnes Electrical Power Engineer's Handbook. 2nd ed. Oxford: Elsevier Publications. 2005

[50] Jahanbin T, Hashim M, Matori KA. Comparative studies on the structure and electromagnetic properties of Ni–Zn ferrites prepared via co-precipitation and conventional ceramic processing routes. Journal of Magnetism and Magnetic Materials. 2010;**322**(18):2684-2689. DOI: http://dx.doi.org/10.1016/j.jmmm.2010.04.008

Chapter 5

Sintering of Whiteware Body Depending on Different Fluxing Agents and Binders

Radomir Sokolar

Additional information is available at the end of the chapter

http://dx.doi.org/10.5772/68082

Abstract

The sintering of whiteware (porcelain) body can be affected by using fluxing agents or binders. The chapter describes the sintering process of porcelain body in case of different fluxing agent (different feldspar rocks, bone ash, zeolite) and binder (kaolin vs. calcium aluminate cement) utilization in the porcelain raw material mixture. Sintering process is presented according to thermodilatometrical curves and sintering temperatures especially.

Keywords: whitewares, feldspar rocks, zeolite, bone ash, kaolin, calcium aluminate cement, sintering temperature, water absorption, mineralogical composition

1. Introduction

Whiteware is a traditional ceramic material used to make pottery and porcelain. Traditional raw material mixture for whiteware (porcelain) production covers kaolin or/and kaolin clay, quartz and feldspar rock at a composition about 50:25:25 wt.%. Typical properties of porcelain body are low porosity (below 0.3%), high mechanical strength (bending strength over 40 MPa, Young Modulus over 60 GPa), firing temperature about 1300°C and high whiteness and translucency [1–3].

Feldspar rocks are used in the fine ceramic industry as a fluxing agent to form a glassy phase for accelerating of sintering process. Feldspar rocks are a mixture of pure feldspars, quartz and mica especially from the mineralogical point of view. Pure feldspars are divided into potassium feldspars (orthoclase, microcline), sodium feldspars (albite) and calcium feldspars (anorthite). Solid solutions between K-feldspar and albite are called alkali feldspars, and solid solutions between albite and anorthite are plagioclase feldspars.

The plagioclase series follows according to percentage of anorthite in parentheses [4]. Feldspar rocks are usually used as a source of alkali oxides (Na_2O, K_2O) and alumina (Al_2O_3) for the preparation of glazes [5]. Suitable choice of feldspar rock can significantly affect the properties of the ceramic body [6], firing temperature and soaking time [10]. The densification of green body, cleanability and the stain resistance of polished sintered ceramic tiles is influenced by particle size distribution of used feldspar rocks [7]. Feldspar rocks may be successfully replaced by LCD waste glass [8]. Wollastonite is very suitable material for acceleration of sintering process in porcelain body. Only 1 wt.% addition of wollastonite is able to decrease firing temperature (about 25°C) in the mixture with kaolin, quartz and potassium feldspar rock [9].

Bone ash is fluxing agent for artistic porcelain especially known as bone china. The amount of bone ash in the raw material mixture of bone china is about 50% [11]. Bone ash (cattle bones calcined at around 1000°C) consists predominantly of hydroxyapatite. The reactions of bone ash in porcelain body were studied in detail in Refs. [12, 13]. Bone ash—fluxing agent for bone porcelains (bone china)—is usually produced by the calcination of bovine bones at the temperature of 1100°C. The melting point of bone ash is about 1670°C [14]. The mineralogical composition of bone ash consists of tricalcium phosphate in the form of hydroxyapatite $Ca_5(OH)(PO_4)_3$.

Very useful fluxing agent for sintered ceramic body production is zeolite, which is able to accelerate the sintering process very intensively. Zeolite is a natural mineral with exceptional physical properties that follow from its specific crystal structure. The latter consists of a 3D lattice of silicate tetrahedrons $(SiO_4)^{4-}$ mutually connected by oxygen atoms, with part of silicon atoms replaced by aluminum atoms $(AlO_4)^{5-}$. Zeolite has a wide range of applications in agriculture, breeding, civil engineering, protection of environment, wastewater purification, and in various industrial sectors. In civil engineering, it began to be used as a partial replacement of cement in the production of concrete [15–19]. Different Italian low-cost natural zeolitic rocks as a substitute of feldspar rocks in porcelain raw materials mixture were investigated. Zeolitic rocks increased the slip viscosity during wet grinding with a coarser grain size distribution. The technological properties (strength, porosity, resistance etc.) of zeolite-based porcelain bodies are similar to current traditional porcelain bodies made in the system kaolin—feldspar rock—quartz [20]. The aim of the study [21] was to investigate the effect of natural zeolite addition on the sintering kinetics. Clinoptilolite, which is a type of natural zeolite, was added partially or fully in replacement of quartz at selected electro-porcelain composition. It was found that the sintering activation energy decreased with increasing zeolite addition. Replacement of quartz with zeolite decreases activation energy for the start of sintering process in electro-porcelain body—firing temperature (about 50–100°C) and soaking time were reduced. In the study [22], the effect of natural zeolite addition on the electrical properties of porcelain bodies was investigated. The resistivity of samples increased at 50°C temperature after zeolite addition, while it was decreasing after zeolite addition at higher temperatures. The resistivity of samples depends on sintering temperature. Low-cost naturally occurring mixtures of feldspar and zeolite occurring in epiclastic rocks were promising substitutes for conventional quartz-feldspathic fluxes in ceramic bodies. Different epiclastic outcrops, with a different zeolite-to-feldspar ratio, were tested in porcelain stoneware bodies. The addition of an epiclastic rock (20 wt.%) brought significant advantages (better grind ability, lower firing temperature with improved mechanical strength and lower porosity) and disadvantages (increasing of slip viscosity, worse powder compressibility, higher firing shrinkage, and a darker color of the body due to high amounts of Fe_2O_3) [23, 24].

Anorthite type of whiteware body on the basis of raw materials mixture of feldspar rock, quartz and calcium aluminate cement (CAC) was developed at the firing temperature of 1300°C. Calcium aluminate cement (substitution of traditional kaolin or quartz) increases the strength of green body (**Figure 1**) and lowers the density due to formation of anorthite in all the fired bodies. An optimal ratio between quartz and feldspar rock for optimal sintering of the body was found (**Figure 2**) [25].

Whiteware body based on anorthite was developed from the mixture of ball clay, alumina, quartz, wollastonite and magnesia mixture. Sintered whiteware body (1220°C) has approximately two times higher modulus of rupture (110 MPa) than traditional porcelain body based on mullite due to lower content of glassy phase (only 30% for anorthitic whiteware body) [26]. Deflocculation of raw materials mixture based on calcium aluminate cement for the production of whiteware body with low porosity is necessary [27]. Carboxylic acids [28], polyethylene glycol, polyacrylate derivatives and aqueous solutions of sodium carboxylate [29] for optimalization of rheological properties of aluminous cement pastes were tested.

Direct sintering is very effective method how to decrease the energy consumption during the firing of porcelain. Direct sintering reduced total processing time by ~50% and also lowered the sintering temperature from 1200 to 1175°C [30].

For the description of sintering process, the thermodilatometrical analysis and sintering temperature are used primarily. Sintering temperature is defined as temperature when the fired body has water absorption exactly 2%.

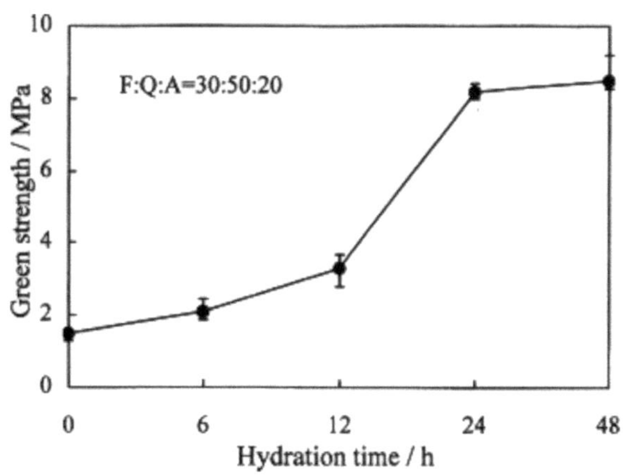

Figure 1. Variation of flexural strength of green body with hydration time. F, feldspar; Q, quartz; A, calcium aluminate cement (CAC) [25].

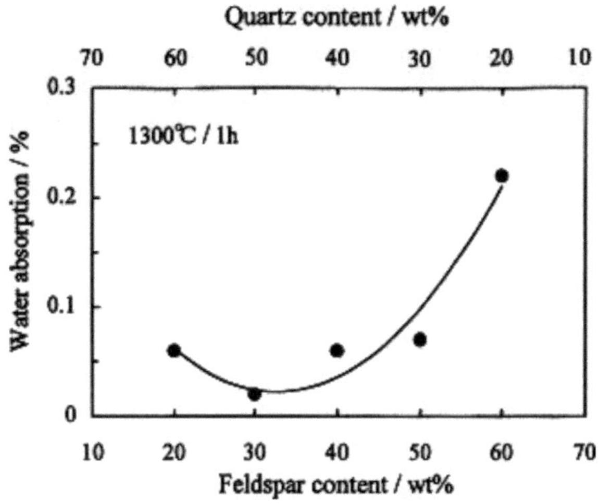

Figure 2. Water adsorption with various compositions in fired bodies with 20% of CAC [25].

2. Sintering of whiteware body depending on fluxing agent (feldspar rocks, bone ash, zeolite)

Sintering and melting of feldspar rocks depend on many aspects, such as the fineness of milling (granulometry), the rate of heating and finally the content of alkali oxides, because it directly creates the melting effect. Very useful is to compare the sintering activity of different typical feldspar rocks with different content of pure K-feldspar, Na-feldspar and Ca-feldspar using for the industrial production of whitewares. The comparison is performed for pure feldspar rocks and for mixtures of feldspar rocks with kaolin. For the comparison, next feldspar rocks were used:

- Sodium-potassium feldspar rock F-KNa with mineralogical composition: K-feldspar (microcline) 20.0%, Na-feldspar (albite) 22.6%, Ca-feldspar (anorthite) 2.4% and quartz 55.0%.
- Potassium feldspar rock F-K with mineralogical composition: K-feldspar (microcline) 57.2%, Na-feldspar (albite) 16.0%, Ca-feldspar (anorthite) 1.5%, quartz 21.3% and mica (muscovite) 4.0%.
- Sodium-calcium feldspar rock F-NaCa with mineralogical composition: Na-feldspar (albite) 60.3%, Ca-feldspar (anorthite) 21.0%, quartz 13.8% and mica (muscovite) 4.9%.

The chemical composition of compared feldspar rocks (**Table 1**) reflects their mineralogical composition and volume of different types of pure feldspars (microcline, albite, anorthite). Granulometry of industrially milled feldspar rocks (the equivalent mean spherical diameter of particles *d*(0.5) in **Table 1**) is very similar and does not affect the presented results.

Sintering activity of dry pressed test samples based on tested pure feldspar rocks (**Table 1**) was determined according to dependence of water absorption (EN ISO 10545) on the firing

	F-KNa	F-K	F-NaCa	Zeolite
SiO_2	79.76	70.96	66.67	68.20
Al_2O_3	12.37	16.10	20.11	12.40
Fe_2O_3	0.42	0.10	0.26	1.40
TiO_2	0.05	0.04	0.04	–
CaO	0.48	0.30	4.23	3.30
MgO	0.10	0.06	0.07	1.00
K_2O	3.35	10.36	0.83	2.80
Na_2O	2.67	1.90	7.13	1.00
LOI	0.80	0.20	0.74	–
d(0.5) [μm]	20.8	18.4	16.6	20.0

Table 1. Chemical composition of used feldspar rocks and zeolite in weight% (LOI = loss of ignition) and the equivalent mean spherical diameter *d*(0.5).

temperature (**Figure 3**). The most intensive sintering activity of the pure feldspar rock body shows potassium-sodium feldspar rock F-KNa—dry pressed test samples have the lowest water absorption, the highest bulk density and modulus of rupture in all firing temperatures in the range of firing at temperatures 1120–1210°C. Sodium-calcium feldspar rock F-NaCa begins sintering at much higher firing temperatures. Sintering temperature (**Figure 3**) of tested alkali feldspar rocks F-KNa and F-K is significantly lower than oligoclase type of feldspar rock F-NaCa.

The mixtures of feldspar rocks with kaolin (40 wt.%)—samples FK-KNa, FK-K, FK-NaCa—totally change (increase) the sintering temperatures (**Table 2**) of alkali feldspar rocks F-K and F-KNa. The most intensive fluxing agent in case of pure feldspar rock body (F-KNa) exhibits the lowest sintering activity in the mixture with kaolin with the highest sintering temperature. This fact is confirmed according to thermodilatometrical curves (**Figure 4**). Conversely, the mixture with kaolin decreases the sintering temperature of oligoclase F-NaCa with the highest content of pure feldspars. The sintering temperature of F-NaCa mixture with kaolin is lower (about 20°C) than pure feldspar rock F-NaCa.

The difference between the sintering of pure feldspar rocks (F-KNa, F-K, F-NaCa) and mixtures of feldspar rocks with kaolin (FK-KNa, FK-K, FK-NaCa) is evident from the thermodilatometrical curves (**Figure 4**). The highest content of quartz and muscovite in feldspar rock F-KNa caused high expansion of the body during firing in the range of 200–900°C in comparison with other tested samples based on feldspar rocks F-NaCa and F-K. Dry pressed body based on pure feldspar rock F-KNa shows the best sinterability of all compared feldspar rocks with maximal firing shrinkage (about 5%—**Figure 4**). Very significant is quartz transformation at the temperature 573°C on cooling part of thermodilatometrical curves (**Figure 4**) depending on quartz content (**Table 1**) in individual tested feldspar rocks. The quartz transformation is most visible for F-KNa feldspar rock with maximal (55%) content of quartz.

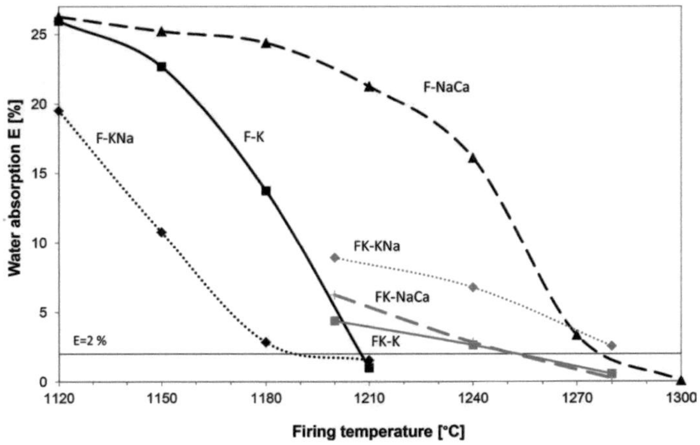

Figure 3. Water absorption *E* depending on the firing temperature. Determination of sintering temperature ($E = 2\%$).

Mixture	Sintering temperature (°C)	Mixture	Sintering temperature (°C)
F-K	1205	FK-K	1250 (+50)
F-NaCa	1275	FK-NaCa	1255 (−20)
F-KNa	1190	FK-KNa	1285 (+95)
		FK-B	1200

Table 2. Sintering temperatures of tested samples based on different feldspar rocks and mixtures of kaolin (60%) with feldspar rocks (40%) or bone ash (FK-B).

The feldspar rock F-KNa with the lowest sintering temperature based on microcline and albite is typical by the quickest disappearing of feldspars during the sintering. Sintering temperature (1190°C) means the existence of only quartz and amorphous glassy phase without any feldspars (**Figure 5**). Quartz, amorphous glassy phase, and microcline are represented in the body F-K after the firing at sintering temperature (1205°C). It is not possible to find an explanation of this fact in granulometry parameters of used feldspar rocks, which influence sintering and melting of feldspars very much, but in the equilibrium phase diagrams (**Figure 6**). Mixed sodium-potassium feldspar rock generated low melting eutectic melts, which accelerate the sintering and melting process of feldspars. It is surprising that leucite generating during the potassium feldspars melting according to theoretical assumptions [4] is not detected even in sintered body F-KNa or sintered body F-K, both based on the potassium feldspar microcline. After the firing of F-NaCa sample at sintering temperature (1275°C), the body contains anorthite (calcium feldspar) with high theoretical melting temperature of about 1550°C [4] and albite (**Figure 5**).

The more intensive fluxing agent than feldspar rocks for the sintering process in the system kaolin-fluxing agent is bone ash (**Figure 4**)—the mixture containing bone ash FK-B (**Table 2**) shows sintering temperature 1200°C. That is about 50°C lower compared with the most intensive feldspar rock-based mixture (FK-K) with potassium feldspar rock F-K containing 75% of pure microcline (**Table 2**). After the exceeding, the temperature 1200°C is visible intensive

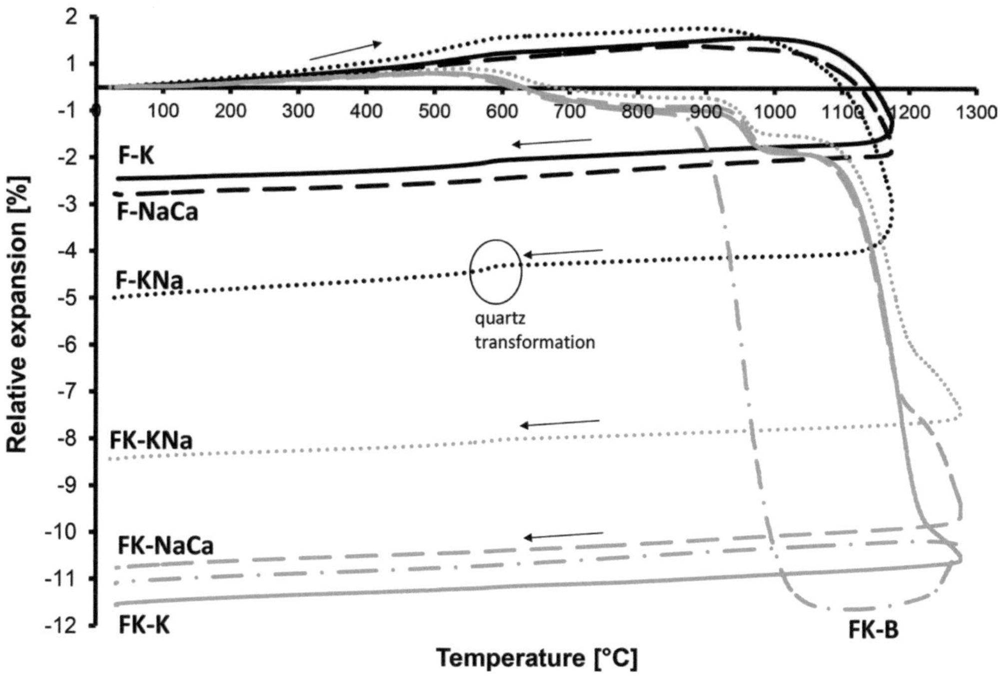

Figure 4. Thermodilatometric curves of pure feldspar rocks (F-K, F-KNa, F-NaCa) and the mixtures of feldspar rocks with kaolin (FK-K, FK-KNa, FK-NaCa) (10°C/min without soaking time on the maximal temperature).

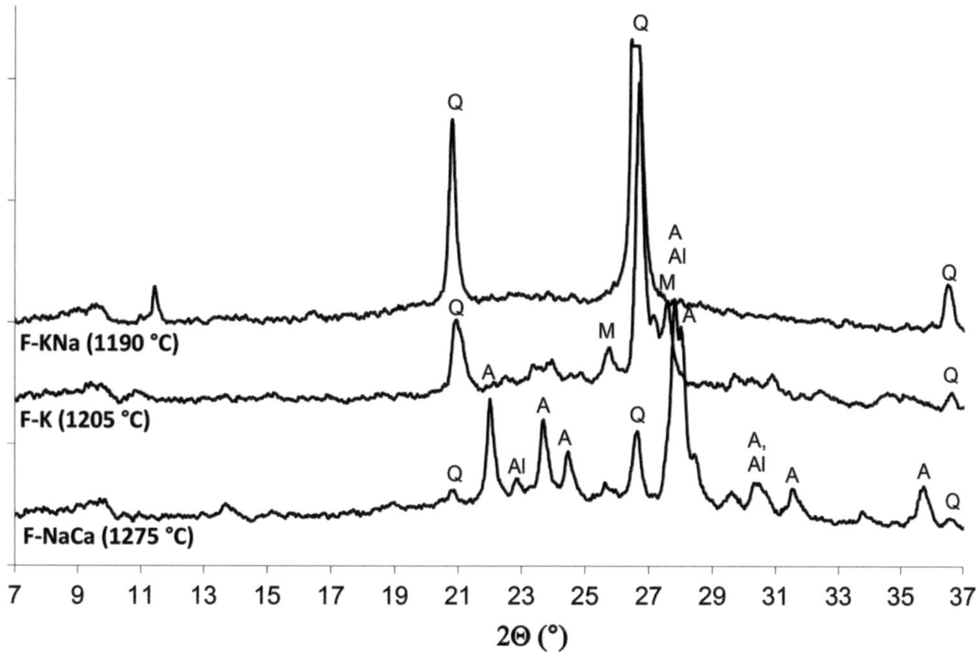

Figure 5. XRD patterns of sintered feldspar rocks at sintering temperature: M, microcline; Al, albite; Q, quartz; A, anorthite.

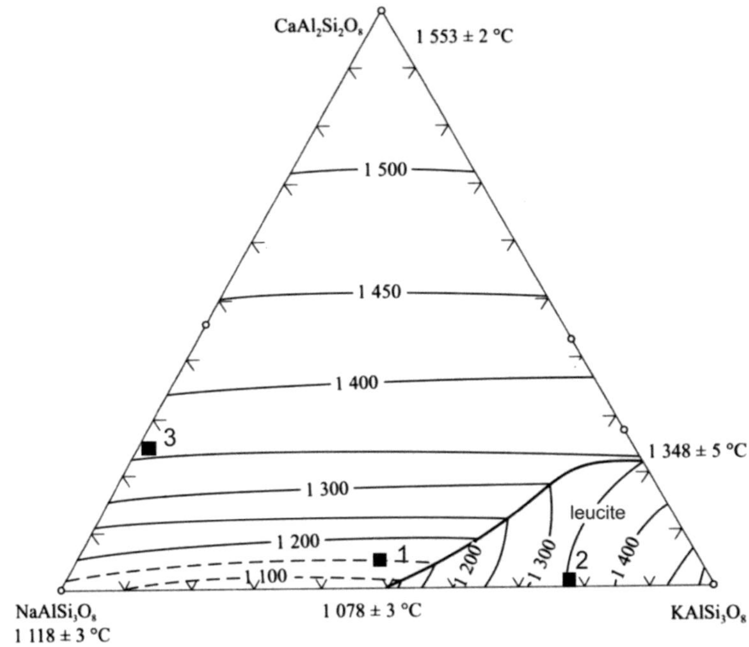

Figure 6. Phase diagram $NaAlSi_3O_8$—$KalSi_3O_8$—$CaAl_2Si_2O_8$ and theoretical melting temperature of used feldspar rocks (1: F-KNa, 2: F-K, 3: F-NaCa).

bloating of the bone ash bodies, which is typical by creating of secondary porosity and increasing in water absorption (**Figure 4**).

Different mineralogical composition between feldspar rocks and bone ash-based porcelain sintered bodies is possible to document according to XRD analyses. Traditional porcelain with high content of feldspar rocks in the raw materials mixture contains mullite and quartz as main mineralogical phases. Mineralogical composition of porcelain body based on bone ash is totally different—typical is high content of β-tricalcium phosphate and anorthite. Bone ash in bone porcelain bodies decomposes into β-tricalcium phosphate $Ca_3(PO_4)_2$, lime CaO and water at around 775°C according to Eq. (1) [14]:

$$Ca_{10}(PO_4)_6(OH)_2 \rightarrow 3\,\beta\text{-}Ca_3(PO_4)_2 + CaO + H_2O \tag{1}$$

Lime reacts with metakaolin from clay relicts to form of anorthite [$CaAl_2Si_2O_8$] according to Eq. (2) [14]:

$$Al_2O_3 \cdot 2\,SiO_2 + CaO \rightarrow CaAl_2Si_2O_8 \tag{2}$$

Eutectic composition in the ternary system of bone china ($Ca_3(PO_4)_2$—$CaAl_2Si_2O_8$—SiO_2) is about 11% tricalcium phosphate, 51% anorthite and 38% silica with a melting temperature of 1290 ± 5°C [14].

Zeolite rock was investigated as a fluxing agent for sintered ceramic body and its effect in the sintering process. The thermodilatometric heating and cooling curves dL/L0 of two different samples according to fluxing agent utilization (zeolite vs. feldspar rock F-KNa) are shown in **Figure 7**. During the firing, there is evident (**Figure 7**) that zeolite (in mixture Z) is more intensive fluxing agent compared to feldspar rock F-KNa (mixture F) for the creation of sintered body with low porosity. High firing shrinkage is typical for the sintering—the raw materials mixture not C, but Z with zeolite content starts intensive shrinking from temperature of about 900°C. Compared mixture F based on traditional ceramic fluxin agent - potassium feldspar rock F-K - starts the sintering process at a higher temperature (about 1100 °C).

The quartz transformation at 573°C (change of the fired body volume) is visible on cooling part of thermodilatometric curve of the mixture F (**Figure 7**) due to high portion of quartz in the mixture F based on the mixture kaolin-quartz-feldspar. This phenomenon is not presented on the cooling curve of the sintered body Z based on zeolite—the raw material mixture not contains quartz, which is advantageous for lower relative expansion (coefficient of linear thermal expansion) of sintered body Z (**Figure 7**).

Sintering temperature of tested samples based on zeolite is 1180°C, which is about 100°C lower than for the sample based on standard flux feldspar rock F-K (mixture F). From the picture (**Figure 7**), there is evident different coefficient of linear thermal expansion α (in the temperature range of 30–500°C) of both compared sintered bodies (Z vs. F):

- feldspar + quartz body: $\alpha_{30\text{–}500°C}$ (F) = 70 × 10^{-7} K^{-1}
- zeolite body: $\alpha_{30\text{–}500°C}$ (Z) = 48 × 10^{-7} K^{-1}

The sintered body based on zeolite shows the lower coefficient of thermal expansion α compared with feldspar sample due to the formation of anorthite in the sample Z and absence of quartz.

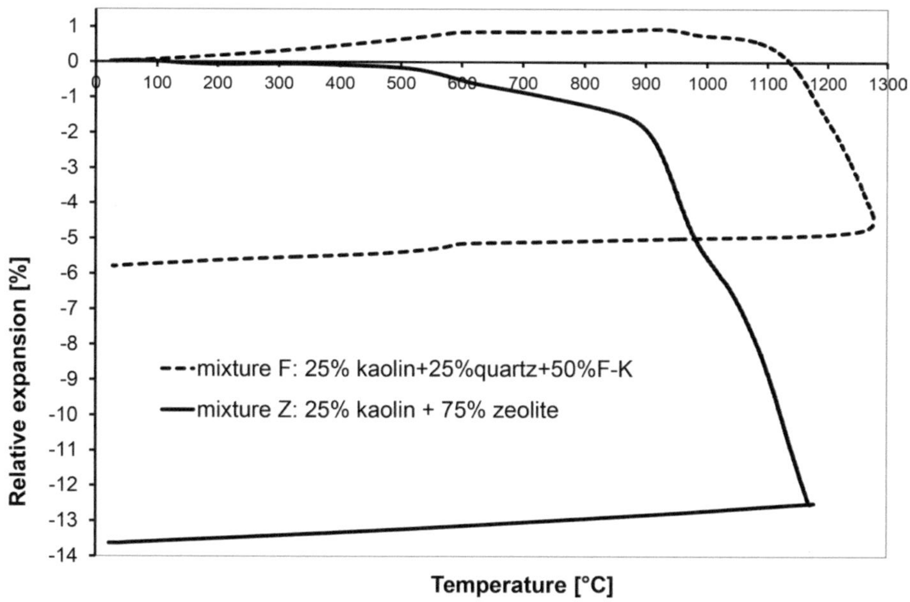

Figure 7. Thermodilatometric analysis of compared samples with different fluxing agent (zeolite vs. feldspar rock F-KNa) during the firing (5°C/min without soaking time).

Important technical property of anorthite is its low coefficient of linear thermal expansion of 48.2 × 10^{-6} K^{-1}[31] (mullite 60 × 10^{-7} K^{-1} [32]). The mineralogical composition of both bodies after firing in both cases is characterized by the existence of mullite and glass phase. The sintered body (fired at 1200°C—mixture Z or 1300°C—mixture F, respectively) based on feldspar and quartz (mixture F) also contains quartz, and the body made from zeolite contains anorthite and cristobalite.

Sintered body based on zeolite (mixture Z in **Figure 8**) as a fluxing agent not creates white body, which is typical for sintered body of the mixture F based on F-KNa feldspar rock (**Figure 8**). This situation corresponds to chemical composition of natural zeolite with higher content of Fe_2O_3 (**Table 1**).

Figure 8. Color of sintered bodies with water absorption below 2%.

3. The effect of calcium aluminate cement as a binder for the sintering of whiteware bodies

The sintering process of whiteware (porcelain) body is affected by the used binder—we can use traditional plastic material (kaolin) or calcium aluminate cement (CAC) according to latest research [25, 33]. Comparison of the properties of both types (**Table 3**) of porcelains made by pressing from dry granulate is documented.

The difference in sintering process of two whiteware bodies with different binder kaolin vs. CAC (**Table 3**) is documented according to thermodilatometric curves (**Figure 9**). The sintering activity of both compared mixtures is very different when the firing temperature exceeds 1200°C—the system based on CAC (mixture CAC) is more able to sinter—we can observe higher firing shrinkage.

Significant decrease of the coefficient of linear thermal expansion in the temperature range 30–500°C is evident (**Figure 9**) when calcium aluminate cement CAC is used as binder compared with kaolin based body. The explanation of this fact we can find in the formation of anorthite in the CAC-based sample (**Figure 10**). The fired body based on kaolin also contains mullite and quartz as a main mineralogical phases. Anorthite exhibits lower coefficient of linear thermal expansion of 48.2 × 10^{-6} K^{-1} [31] than mullite 60 × 10^{-6} K^{-1} [32].

Mixture	Content (%-mass)
K	25% kaolin + 50% F-KNa + 25% quartz sand + 0.35% sodium hexametaphosphate (deflocculant)
CAC	25% CAC + 50% F-KNa + 25% quartz sand + 0.35% sodium hexametaphosphate (deflocculant)

Table 3. Composition of raw material mixtures (test samples).

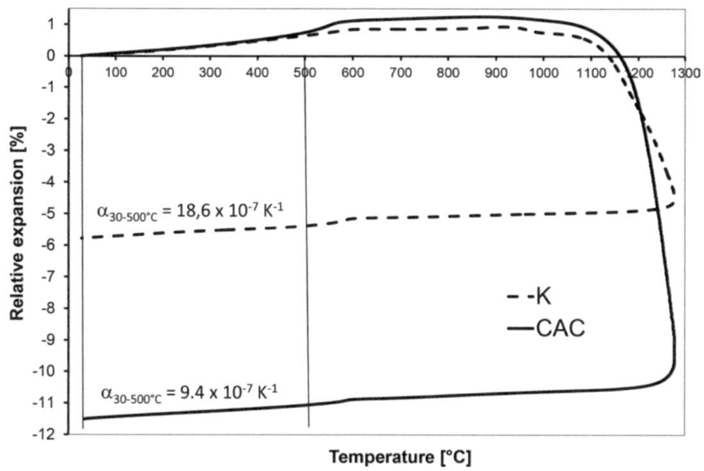

Figure 9. Thermodilatometric analysis of kaolin (K) and calcium aluminate cement (CAC)-based bodies during the firing (1280°C, 3°C/min without soaking time). Determination of the coefficient of linear thermal expansion in the range of temperatures 30–500°C.

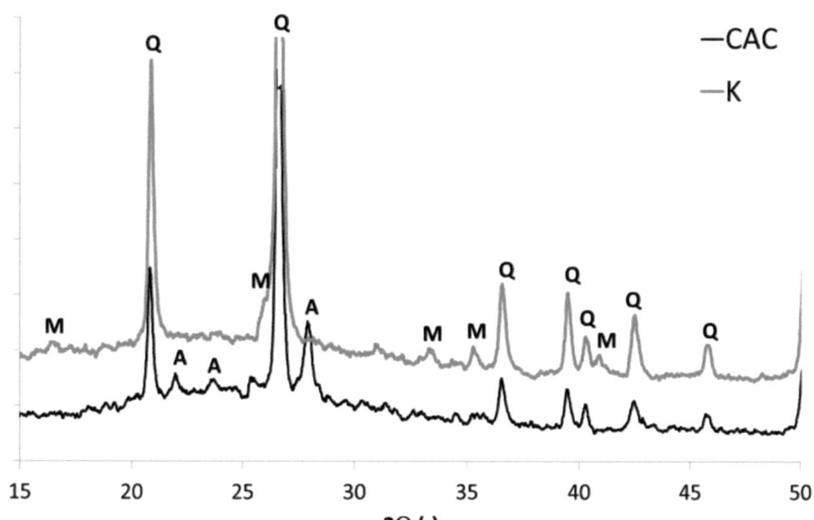

Figure 10. XRD of fired bodies based on different binder kaolin or CAC (M, mullite; Q, quartz; A, anorthite).

Sample	Firing temperature (°C)					
	1250			1280		
	Ev (%)	MOR (MPa)	B (kg m^{-3}]	Ev (%)	MOR (MPa)	B (kg m^{-3}]
K	6.4	29.4	2180	1.9	38.5	2280
CAC	1.3	58.9	2360	Melting of test samples		

Table 4. Physicomechanical properties of fired bodies K and CAC depending on firing temperature: Ev—water absorption, MOR—modulus of rupture, B—bulk density.

The CAC mixture shows significantly higher sintering activity according to measured parameters of porosity—the prepared samples of a mixture CAC have a lower water absorption Ev and higher bulk density B (according to EN ISO 10545) than mixtures based on kaolin (K) after firing at the same temperature (**Table 4**). Higher modulus of rupture MOR of fired bodies is achieved for anorthitic type of body (CAC) compared with mullite whiteware body (K) when MOR values for samples with similar porosity are compared (K-1280°C and CAC-1250°C in **Table 4**). Similar results are published in Ref. [25].

4. Conclusions

For sintering and melting of pure feldspar rocks, not just the total content of feldspar components is important, but also the ratio between potassium, sodium and calcium feldspars. At the appropriate ratio, low melting eutectics can be expected to rise, with a melting temperature substantially lower than the theoretical melting temperature of pure feldspars. The presence of calcium feldspar significantly reduces sintering ability and melting of feldspar rocks. Totally different results we can expect for the mixtures of feldspar rocks with the plastic part of whiteware raw materials mixture—kaolin. The reactions between feldspar rocks and kaolin (Al_2O_3, SiO_2) during the sintering process are the cause of low melting eutectics, which accelerate sintering.

Natural zeolite is very intensive fluxing agent for ceramic technology. Using zeolite we can reduce the sintering temperature of the body of about 100°C, compared with traditional ceramic fluxing agent—potassium-sodium feldspar rock F-KNa. The sintered body (with water absorption below 2%) based on zeolite has lower coefficient of linear thermal expansion. The presence of zeolite in raw materials mixture significantly changes mineralogical composition of fired whiteware body—mullite, anorthite and cristobalite are the main mineralogical phases instead of mullite and quartz, which are typical for a standard whiteware bodies made from raw material mixtures based on kaolin, quartz and feldspar. The limiting factor for the use of natural zeolite as a flux for whiteware is its coloring effect.

Calcium aluminate cement CAC with high content of Al_2O_3 (70%) in the raw materials mixture for whiteware production is suitable alternative to kaolin—higher strength of green and fired body, more intensive whiteness of body after firing and lower coefficient of linear thermal expansion is possible to expect using CAC. The sintering activity of the whiteware body is accelerated when calcium aluminate cement is used as a binder instead of kaolin—the bodies can be fired at lower temperatures. Calcium aluminate cement significantly changes mineralogical composition of fired body—anorthite is the main mineralogical phase, mullite is typical phase for standard porcelain bodies made in the system of kaolin-quartz-feldspar rock.

Acknowledgements

The results were achieved under the project no. LO1408 "AdMaS UP—Advanced Materials, Structures and Technologies", supported by the Ministry of Education, Youth and Sports under the "National Sustainability Programme I" (chapter 2). The results were achieved under the project—the Czech Science Foundation, research project no. P104/13/23051S "Anorthite porcelain body on the basis of aluminous cement" (chapter 3).

Author details

Radomir Sokolar

Address all correspondence to: sokolar.r@fce.vutbr.cz

Faculty of Civil Engineering, Brno University of Technology, Brno, Czech Republic

References

[1] Carty WM, Senapati U. Porcelain—raw materials, processing, phase evolution, and mechanical behavior. Journal of American Ceramic Society. 1998;**81**:3-20

[2] Rado P. An Introduction to the Technology of Pottery. Oxford: Pergamon Press; 1988

[3] Stubna I, Slavikova J, Vozar L. Relationship between mechanical strength and Young's modulus of porcelain. Industrial Ceramics, 2008;**28**(2):153-154

[4] Barth TFW. Feldspars. 1st ed. Bath: John Wiley & Sons; 1969

[5] Norton, FH. Fine Ceramics Technology and Applications. 1st ed. Malabar: R.E Krieger; 1970

[6] Das S. Kr, Dana K. Differences in densification behaviour of K- and Na-feldspar-containing porcelain bodies. Thermochimica Acta, 2003;**406**(1-2):199-206

[7] Alves HJ, Melchiades FG, Boschi AO. Effect of feldspar particle size on the porous microstructure and stain resistance of polished porcelain tiles. Journal of the European Ceramic Society, 2012;**32**(10):2095-2102

[8] Kim K, Kim K, Hwang J. LCD waste glass as a substitute for feldspar in the porcelain sanitary ware production. Ceramics International. 2015;**41**(5):7097-7102

[9] Turkmen O, Kucuk A, Akpinar S. Effect of wollastonite addition on sintering of hard porcelain. Ceramics International. 2015;**41**(4):5505-5512

[10] Sokolar R, Vodova L. Sintering behaviour of feldspar rocks. Research Inventy: International Journal of Engineering and Science. 2014;**10:**49-55

[11] Sokolar, Vodova, L. The difference between traditional and bone porcelain body. Advanced Materials Research. 2015;**1100**:87-90

[12] Iqbal Y, Messer PF, Lee WE. Microstructural evolution in bone China. British Ceramic Transactions. 2000;**99**:193-199

[13] Iqbal Y, Messer PF, Lee WE. Non-equilibrium microstructure of bone China. British Ceramic Transactions. 2000;**99**:110-116

[14] St Pierre JPDS. Note on the System CaO-Al_2O_3-P_2O_5. Journal of the American Ceramic Society. 1956;**39**:147

[15] Poon CS, Lam L, Kou SC, Lin ZS. A study on the hydration rate of natural zeolite blended cement pastes. Construction and Building Materials. 1999;**13**(1999):427-432

[16] Valipour M, Pargar F, Shekarchi M, Khani S. Comparing a natural pozzolan, zeolite, to metakaolin and silica fume in terms of their effect on the durability characteristics of concrete: a laboratory study. Construction and Building Materials. 2013;**41**(2013):879-888

[17] Canpolat F, Yilmaz K, Kose MM, Sumer M, Yurdusev MA. Use of zeolite, coal bottom ash and fly ash as replacement materials in cement production. Cement and Concrete Research. 2004;**34**(2004):731-735

[18] Vejmelková E, Keppert M, Ondráček M, Černý R. Effect of natural zeolite on the properties of high performance concrete. Cement Wapno Beton. 2013;**18**(2013):150-159

[19] Ranjbar MM, Madandoust R, Mousavi SY, Yosefi S. Effects of natural zeolite on the fresh and hardened properties of self-compacted concrete. Construction and Building Materials. 2013;**47**(2013):806-813

[20] de Gennaro R, Cappelletti P, Cerri G, de Gennaro M, Dondi M, Guarini G, Langella A, Naimo D. Influence of zeolites on the sintering and technological properties of porcelain stoneware tiles. Journal of the European Ceramic Society. 2003;**23**(2003):2237-2245

[21] Demirkirana AS, Artirb R, Avcia E. Effect of natural zeolite addition on sintering kinetics of porcelain bodies. Journal of Materials Processing Technology. 2008;**203**(2008):465-470

[22] Sukran Demirkiran A, Artir R, Avci E. Electrical resistivity of porcelain bodies with natural zeolite addition. Ceramics International. 2010;**36**(2010):917-921

[23] de Gennaro R, Dondi M, Cappelletti P, Cerri G, de' Gennaro M, Guarini G, Langella A, Parlato L, Zanelli Ch. Zeolite–feldspar epiclastic rocks as flux in ceramic tile manufacturing. Microporous and Mesoporous Materials. 2007;**105**(2007):273-278

[24] Sokolar R, Sveda M. The use of zeolite as fluxing agent for whitewares. Procedia Engineering. 2016;**151**:229-235

[25] Tai W, Kimura K, Jinnai K. A new approach to anorthite porcelain bodies using nonplastic raw materials. Journal of the European Ceramic Society. 2002;**22**(2002):463

[26] Taskiran MU, Demirkol N, Capoglu A. A new porcelainised stoneware material based on anorthite. Journal of the European Ceramic Society. 2005;**25**(2005):293-300

[27] Sokolar R, Vodova L. Sodium Hexametaphosphate as Deflocculation Agent for Calcium Aluminate Cements in Porcelain Body. Advanced Materials Research. **897**:30-33

[28] El Hafiane Y, Smith A, Bonnet JP, Tanouti B. Effect of a carboxylic acid on the rheological behavior of an aluminous cement paste and consequences on the properties of the hardened material. Journal of the European Ceramic Society. 2005;**25**(2005):1143

[29] El Hafiane Y, Smith A, Chartier T, Abouliatim Y, Nibou L, Bonnet JP. Role of dispersant and humidity on the setting of millimetric films of aluminous cement prepared by tape casting. Journal of the European Ceramic Society. 2012;**32**(2012):2103

[30] Lerdprom W, Chinnam RK, Jayaseelan DD, Lee WE. Porcelain production by direct sintering. Journal of the European Ceramic Society. 2016;**36**(16):4319-4325

[31] Potuzak M, Solvang M, Dingwell D. Temperature independent thermal expansivities of calcium aluminosilicates melts between 1150 and 1973 K in the system anorthite–wollostanite–gehlenite (An–Wo–Geh): a density model. Geochimica et Cosmochimica Acta. 2006;**70**(2006):3059-3074

[32] Camerucci MA, Urretavizcaya G, Castro MS, Cavalieri AL. Electrical properties and thermal expansion of cordierite and cordierite-mullite materials. Journal of the European Ceramic Society. 2001;**21**(2001):2917-2923

[33] Sokolar R, Vodova L. Whitevare bodies without Kaolin. Interceram. 2014;**63**(2014):19

Section 2

Energy Assisted Sintering

Chapter 6

Sintering and Reactive Sintering by Spark Plasma Sintering (SPS)

Giulia Franceschin, Nancy Flores-Martínez, Gabriela Vázquez-Victorio, Souad Ammar and Raul Valenzuela

Additional information is available at the end of the chapter

http://dx.doi.org/10.5772/intechopen.68871

Abstract

A wide variety of technological applications, especially in electronics, requires high-density nanostructured solids, consolidated by sintering from nanoparticles. A new sintering technique known as spark plasma sintering (SPS) appears as the only method to reach high densities while preserving the final grain size within the nanometric range, with the added advantage of carrying out the process at significantly lower temperatures and shorter times as compared with the classical processes. Recent studies have revealed that in many cases, SPS can also accomplish the solid-state reaction to achieve the desired compound, leading to reactive SPS (RSPS). In this chapter, a review of RSPS is presented, focusing particularly on magnetic oxide materials as functional solids.

Keywords: nanostructured solids, ceramic materials, reactive spark plasma sintering, solid-state processing, magnetic properties

1. Introduction

For the past two decades, the synthesis and applications of magnetic nanoparticles (MNPs) have gained immense interest in a wide range of technologies, especially in the biomedical field [1–4]. These applications are based on the novel magnetic properties associated with the nanoscale [5]. In the electronics field, nanostructured materials point also to innovative applications [6, 7], particularly in magnetic recording, actuators, and microwave devices. For these applications, however, a powder constituted by MNPs is not suitable; a high density, consolidated solid is required. Consolidation by classic sintering methods requires high temperatures and long times—more than 1000°C for several hours—to reach densities above 90%. Such conditions lead to an excessive grain growth, which deteriorate the properties associated with the nanoscale, thus making such consolidation methods impractical.

Spark plasma sintering (SPS) [8] has recently been revealed as an extremely efficient sintering technique for consolidating nanopowders into high density, nanostructured materials. In practice, the powders are heated in a conductive SPS die at very high rates by the action of electric pulses and maintained under uniaxial pressure (**Figure 1**), leading to their sintering with impressive shorter times and lower temperatures than in conventional methods. In addition, to be a more efficient method, it allows a tight control of grain growth, thus permitting the production of nanostructured materials. The principle of SPS and convenient design of the facilities make it attractive for conducting materials. Recent results show, however, that it is equally powerful for nonconductive ones [9].

The current is applied and passed through the graphite die. If the sample is nonconductive, the heat generated inside the walls of the die assists in powder consolidation. If starting powders are a conductive material, the current goes through the powder, and the first sparks are in the surface atoms as well as in surface defects. This punctual warming of atoms is known as hot spots. In these zones, the temperature increases thousands of degrees in a very short time, and nucleation and grain growth begin. If the sample is an insulating phase, the electric field associated with the electric pulses has also a strong effect on atomic diffusion, and sintering is enhanced [9].

SPS has been recently examined in a much broader perspective and has gained a strong reputation as a versatile method of solid-state synthesis, not only for sintering, but also for solid-state reactions, as reported in relevant literature [11, 12]. SPS then becomes reactive SPS (RSPS).

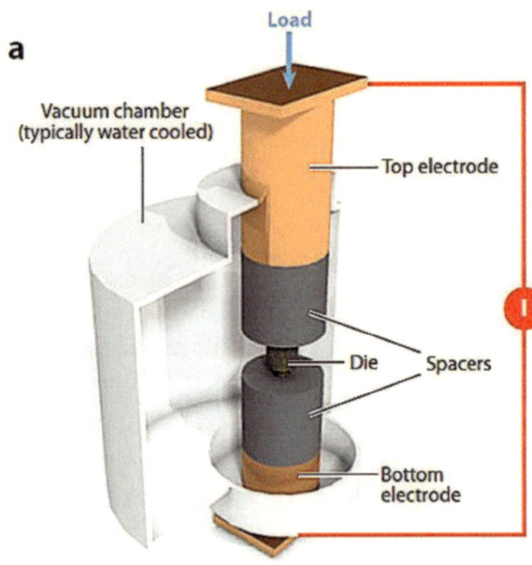

Figure 1. Schematics of the vacuum chamber, electrodes, and pressing die of the SPS system (adapted from Ref. [10]).

In this review, we analyze the microstructure formation of the products of chemical reactions occurring in RSPS, in an attempt to directly produce nanostructured solids starting from the corresponding reactants, that is, an intermediate solid phase, or a mixture of precursors, containing the required elements to form the desired phase. We also discuss the possibility of fabricating nanocomposites, in which the interfaces between the constituting phases can be improved by particular tailoring.

Focusing on magnetic granular oxide nanostructures, we present successful syntheses with a special emphasis on their microstructure stability and attractive properties of the materials. We discuss the challenges of producing a dense nanostructured material when reaction and densification do not coincide during the SPS. Case examples in the fields of magnetocaloric materials (manganites), soft magnetic materials (garnets), and permanent magnets (hexaferrites) are specifically addressed.

We also discuss the limitations of such a technique, in relation to its reducing operating conditions and propose some alternatives to overcome main drawbacks. Indeed, RSPS is mainly performed using graphite-made die and punches under dynamic vacuum, creating a reducing atmosphere. In the case of oxide materials, this can lead to a partial reduction, sometimes even to a metal contamination, affecting the final physical properties of the consolidated solids (electric conductivity, for instance). The replacement of carbon dies by tungsten carbide in such materials offers an interesting alternative.

2. Reduction reactions during SPS processes

The composition of the environment inside the SPS chamber affects the material's diffusivity during sintering. For this reason, the processes that normally occur during a sintering cycle, such as phase constitution, densification, and grain growth, are strongly affected by the sintering atmosphere [13].

Typical materials developed by the SPS technique are refractory metals and intermetallics, oxide and nonoxide ceramics. The particles constituting the powders before consolidation tend to decrease their surface energy by desorption of chemical species, once introduced inside the SPS chamber. The released gas, water, or organic compounds in the atmosphere modify the thermodynamic driving force to surface reduction and sintering.

Commonly, the atmospheres employed for sintering are:

- Vacuum (10^{-4} to 10^{-5} bar),
- Inert gas (up to 1.3 bar), or
- Reducing gas hydrogen-based mixtures.

The atmosphere composition inside the pressing tools which contain the sample may differ considerably from the atmosphere outside [9], making it difficult to control the sintering atmosphere. Sample reduction during RSPS occurs when the thermodynamic conditions are favorable to the imbalance of either one of the following reactions from the right to the left side:

$$\text{metal} + O_2 \leftrightarrows \text{oxide} + \Delta H^0_1 \tag{1}$$

$$\text{metal} + 2H_2O \leftrightarrows \text{oxide} + 2H_2 + \Delta H^0_2 \tag{2}$$

$$\text{metal} + 2CO_2 \leftrightarrows \text{oxide} + 2CO + \Delta H^0_3 \tag{3}$$

ΔH^0_1, ΔH^0_2, and ΔH^0_3 are the heat released per O_2 mole in each oxidizing reaction (from left to right side). By definition, we can calculate the corresponding changes of free enthalpy:

$$\Delta G_1^0 = -\Delta H_1^0 = -RT\ ln\left(\frac{P_{metal}}{P_{oxide}} \cdot P_{O_2}\right) \tag{4}$$

$$\Delta G_2^0 = -\Delta H_2^0 = -RT\ ln\left(\frac{P_{metal}}{P_{oxide}} \cdot \frac{P_{H_2O}^2}{P_{H_2}^2}\right) \tag{5}$$

$$\Delta G_3^0 = -\Delta H_3^0 = -RT\ ln\left(\frac{P_{metal}}{P_{oxide}} \cdot \frac{P_{CO_2}^2}{P_{CO}^2}\right) \tag{6}$$

A standard measure for the tendency of a metal (or a chemical element) to oxidize is given by Eqs. (4)–(6) when $P_{O2} = 1$; thus, we obtain Eq. (7), which is the heat released when 1 mole of O_2 gas at 1 atm pressure combines with 1 mole of metal to form the oxide in function of temperature T:

$$\Delta G^0 = -RT\ ln\left(\frac{P_{metal}}{P_{oxide}}\right) \tag{7}$$

The graphic representation of $\Delta G^0 = f\,(T)$ is known as the Ellingham-Richardson diagram (**Figure 2**). It allows the direct evaluation of the relative affinity between the elements and the oxidizing agent. Elements with a lower line in ΔG^0 in the diagram have a greater affinity to oxygen. This diagram is useful to understand the thermodynamics of the reactions between the sintering material and the atmosphere, and it can give information about the dissociation temperature, the dissociation pressure, and the effect of reducing agents [14].

Special attention must be given to the pressing tools' material. Standard pressing tools used in the RSPS process are graphite based, often internally covered with carbon sheets or foils, in order to ease the removal of the sample after sintering [15]. Thus, graphite components are in close contact with the sample and can become reactive with the oxygen eventually present in the sample itself at temperatures higher than 600°C. Other sources of oxygen are moisture or other gases in the sintering atmosphere. Such a chemical reaction causes the formation of CO and a continuous decrease of the oxygen partial pressure within the furnace, creating a reducing condition in the

http://dx.doi.org/10.5772/intechopen.68871

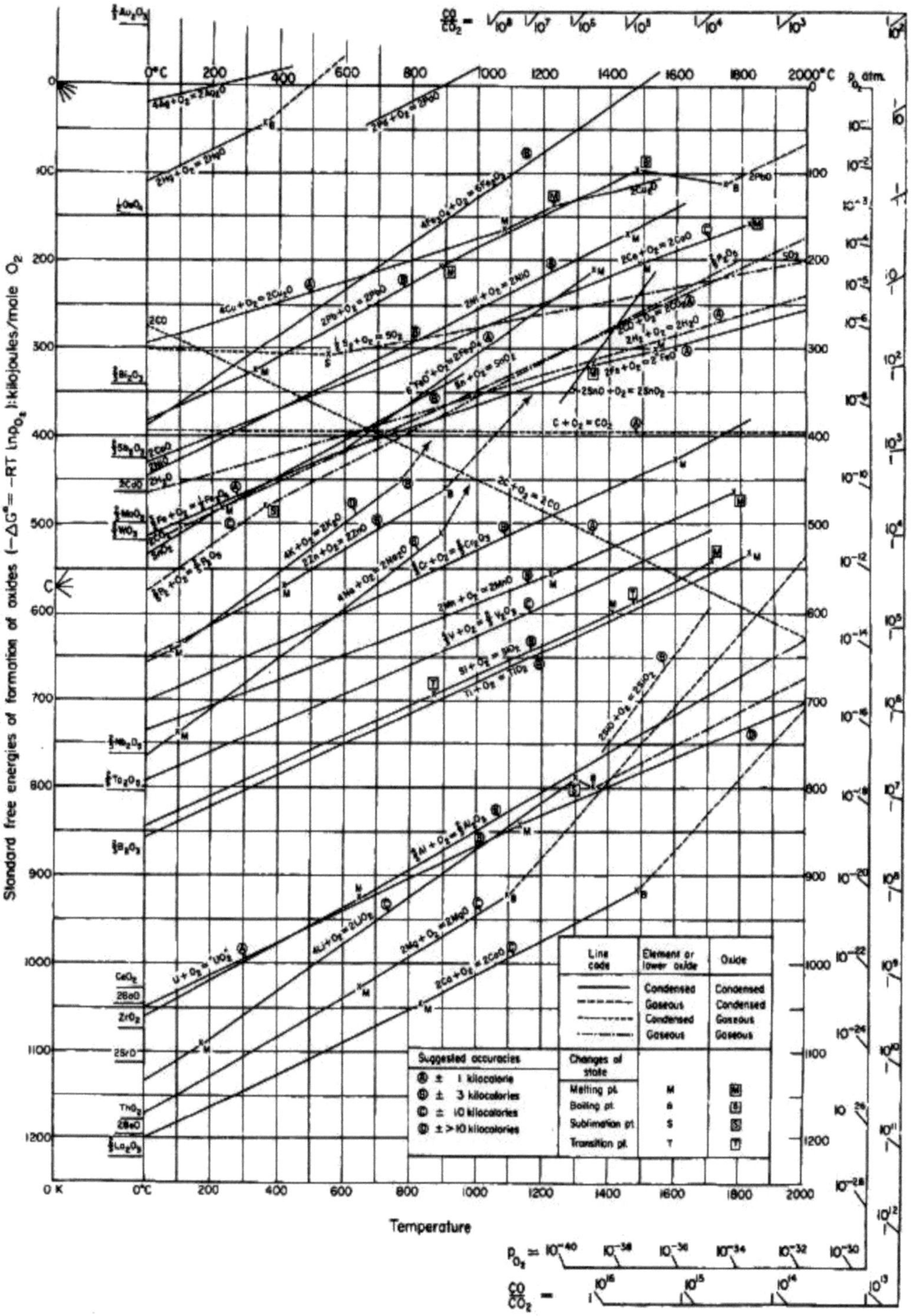

Figure 2. Ellingham-Richardson diagram [16].

sintering atmosphere. When an intense gas phase transport is established between the sample and the mold, reduction of oxides or even precipitation of carbon or carbide in the sample may occur [9].

The materials obtained by the RSPS technique can be distinguished into two main categories in relation to their affinity to reductive atmosphere: (a) if oxide reduction is a desirable effect and (b) if reduction is a secondary effect that should be avoided during the ceramic formation. Metals and nonoxide ceramics constitute the first class of materials, while oxide ceramics represent the second one.

For metals and nonoxide ceramics, reductive atmospheres, such as inert gas or reducing hydrogen gas mixture, are suggested during sintering, because they are effective on cleaning the oxide naturally formed on the metallic surface of the starting powder during air exposition.

The benefit of oxide reduction in this kind of materials is related to the possibility to enhance the sample densification; in fact, oxide compounds possess a smaller density than the corresponding metal, and they hinder atomic diffusion during the densification step.

Such a reductive atmosphere has also been suggested for ultra-high-temperature ceramics (UHTC) to promote their densification. Sometimes, specific additives are mixed to the starting powder to promote reduction and thus enhance densification. For instance, C or B_4C has been used as additives in TaC densification [17], while $MoSi_2$, $TaSi_2$, and SiC have been considered as additives for oxygen removal during HfB_2 sintering [18–20].

Systems requiring a reductive atmosphere during their reactive consolidation are oxide/metal nanocomposites. In practice, the reductive atmosphere can be specifically used for the in-situ formation of metal component. As an example, Al_2O_3/Ni granular solids were produced by reacting and sintering a mixture of Al_2O_3 and NiO powders inside a carbon die [15].

For functional oxide ceramics, including magnetics, reducing sintering atmosphere may have dramatic consequences on the final properties. It may modify the starting oxide composition. Typically, it generates oxygen vacancies, which in the case of transparent ceramics, such as yttrium-aluminum-garnet (YAG), induces light absorption and in-line transmission decrease [21]. In the case of ferromagnetic *p*-doped manganite ceramics, the formed oxygen vacancies decrease the average oxidation state of the paramagnetic manganese cations, inducing a net reduction of the Curie temperature of the final solid in comparison to its conventionally made bulk counterpart [22]. In Ni-Zn ferrite, a well-known resistive soft magnet, consolidation by SPS involves a Fe^{3+} into Fe^{2+}reduction inside the ferrite grains compensated by a loss of Ni cations, which precipitate as Ni metal between the grains, increasing the total electrical conductivity [23].

As a consequence of oxide ceramic changes in reductive atmospheres, solutions to the tools' reactivity at high temperature have been considered. In other words, when the control of the atmosphere is not enough to avoid secondary reduction reactions, new tool materials have been employed. Graphite reinforced with carbon fiber dies [24] is suitable for high-pressure sintering, because of their high mechanical resistance to compression; the possibility to increase the pressure applied to pistons during consolidation allows to operate the process at even lower temperature, thus limiting the possibility of reaction between carbon and oxide inside the material. Tungsten carbide, steel, and refractory metals, such as molybdenum alloys, copper-beryllium, and alumina [25], have been also used as conductive sintering tools. Double-walled tools with inner ceramic die and outer graphite mantle have been also employed [26]. Some works report the use of layers and foils of alumina or other different metals, such as molybdenum, tungsten, and tantalum, which are introduced inside the graphite die to cover the internal mold walls before introducing the sample [9]. By these less-costly operating conditions, the sintering material is never in contact with graphite.

3. Effects of current and pressure

3.1. Electrical current effect

In addition to heating, the effect of current pulses is to enhance mass transport during sintering, more specifically by one of the three mechanisms [27]: (a) increasing the point defect concentration; (b) a reduction in the activation energy for mobility of defects; or (c) electron wind modification of the diffusion flux (electromigration).

Temperature and current are not independent parameters; high heating rates are achieved by increasing pulsed direct current. This is the major difference between the conventional hot-pressing and the SPS methods; in the RSPS method, both the die (typically graphite) and the sample are heated by Joule mechanism from a current passing through them (if the sample is conducting) [27]. Nonconductive materials are heated by means of heat conduction from the die walls. Pulsed direct current can enhance the reaction kinetics when the reactants are brought to interact in the SPS. This effect is, however, system dependent [28]. A change in the electrical conductivity of the materials in the die as reaction progresses can give rise to undesired results. Schmidt et al. [29] studied the decomposition behavior of MgH_2 in RSPS; in order to increase electrical conductivity, graphite was added to the MgH_2 powder. Metallic magnesium, the product of decomposition, increases the electrical conductivity of the material in the SPS die. This example shows that as the reaction product accumulates, the conductivity of the material in the sintering die changes; if it increases, the reaction is self-enhanced due to the presence of in-situ–formed conductive particles inducing the formation of hot spots in the remaining, not yet fully reacted, mixture.

In the case of silica-doped yttria-stabilized zirconia sintering, the electrical resistivity of grain boundaries is often increased by the presence of impurity phases of siliceous compounds. SPS allowed a significant reduction in these compounds while leaving the grains unaffected [31]. This effect was attributed to the generation of electrical discharges between particles as the SPS electrical pulses are applied. The discharges expel the liquid silica phase to triple points in grain boundaries, thus reducing their effects on resistivity.

3.2. Pressure effect

Mechanically, the pressure has a direct effect on particle rearrangement and the destruction of agglomerates, particularly in the case of nanometric powders. However, the significance of the pressure on sintering depends on the particle size. When the particle size is small, the relative contribution of the pressure is small but becomes significant as the particle size increases [27]. In a study on the sintering of nanometric pure zirconia, Skandan et al. [30] found that the pressure had no effect on the relative density of fine-grained powder (6 nm) up to a pressure of about 35 MPa; in contrast, the density increased sharply when higher pressure was used. For larger particle size powder (12 nm), the same behavior was observed except that the transition occurred at about 10 MPa. Another result of the application of pressure is a decrease in the sintering temperature. For the case of SPS densification of a nanometric cubic zirconia, Anselmi-Tamburini et al. [27] showed that the combination of fast heating rate and high pressure produces a marked reduction in the sintering temperature. **Figure 3** shows the effect of pressure on the sintering temperature required to obtain a 95% relative density (with 5-min hold time). The figure also shows the grain size obtained under these conditions. The temperature required to achieve 95% of density decreases linearly with the logarithm of the applied pressure. The grain size varied from about 200 to 15 nm.

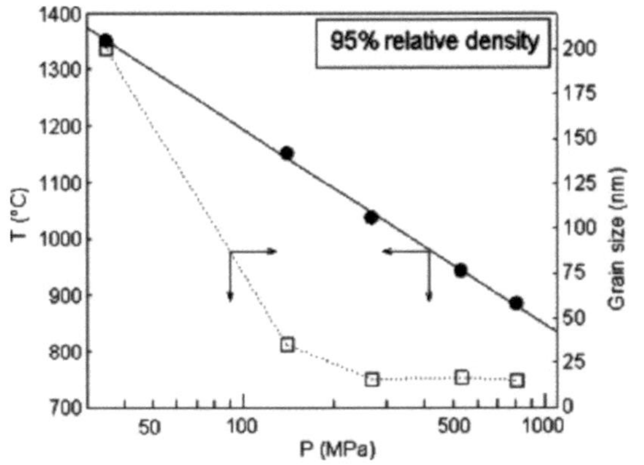

Figure 3. Relationship between hold temperature and the applied pressure required to obtain samples with a relative density of 95% in the case of nanometric fully stabilized zirconia (8% Y_2O_3). Hold time: 5 min. The grain size of the materials is also shown [30].

4. Precursors

4.1. Size of precursors

The size of precursors plays an important role in the final consolidates. In this manner, densification hinges on the characteristics of initial powders inside the die. As mentioned by

Nygren [32], the grain growth is deeply related with the size of starting powder. When nanometric size precursors are employed, most of the driving force to reduce specific area is destined to the densification process.

In the case of micrometric precursors, pressure can enhance densification by four mechanisms [33]: particle rearrangement, localized deformation, bulk deformation, and neck growth. For large particle size of precursors (45–90 μm), density is enhanced through particle rearrangement and localized deformation. In contrast, neck growth increases for smaller size of precursors (~25 μm), while bulk deformation has no influence [33].

Grains can keep the memory of the synthesis process by which they were made, and their characteristics are retained even in the sintered particles [34]. During the processing of Al_2TiO_5 by RSPS, varying the initial powder size and the nature of precursors could enhance consolidation, particularly for sol-gel and co-precipitation synthesis [34]. Co-gelified alumina-titania powders and mechanical mixtures of alumina and titania (both obtained by sol-gel method) and alumina-titania powders (recovered by co-precipitation method) were treated thermally before the RSPS process. The grain sizes for the powders synthesized by sol-gel and co-precipitation method were 10 and 50 nm, respectively. Experimental conditions were the same for all the three samples during RSPS process. The final report showed a smaller increase in the grain size (0.5 μm) for the consolidated powders synthesized by sol-gel as compared with the co-precipitated powders (8 μm). However, the final density for the co-gelified alumina-titania and alumina titania initial powders was very close to the theoretical density (3.7 g/cm^3). Finally, the full phase of Al_2TiO_5 was obtained in co-gelified alumina-synthesized sample.

Different size and aggregation states of polyol-made $CoFe_2O_4$ starting powders were consolidated in similar conditions [35]. In the RSPS process, monodispersed initial powders around 5 nm, and clusters of ~50 nm (made also of ~5 nm particles) were rapidly heated to 600°C for 6 min before rapid cooling. Unexpectedly, the final grain size resulted larger for the monodispersed precursor than for the clustered case. This difference was interpreted on the basis that grain growth is an essentially surface process, and in the monodispersed case, particles offered a larger free surface than in the clustered case.

4.2. Nature of precursors

Preliminary works exploring the possibility of preparing nanostructured manganite ceramics by the RSPS process have evidenced the role of the precursor's nature. Starting from a mixture of raw bulk oxides required a higher reacting and sintering temperature, while starting from a mixture of their hydroxide counterparts allowed a decrease in this operating parameter. Typically, $La_{0,85}Na_{0,15-x}K_xMnO_3$ ceramics were prepared by RSPS starting from the $La(OH)_3$, Na(OH), K(OH), and MnO_2 commercial powder mixture, working at 800°C under a uniaxial pressure of 50 MPs for a couple of minutes (**Figure 4a**) [12]. In contrast, ceramic $LaMnO_3$ was obtained by RSPS starting from La_2O_3 and Mn_2O_3 commercial powder mixture, working at 1000°C under the same pressure for almost the same sintering time (**Figure 4b**) [42]. The final density of both ceramics exceeded 90% of the theoretical value, and their average grain size was in the submicrometer range, the finest grains being obtained at the lowest sintering temperature of course.

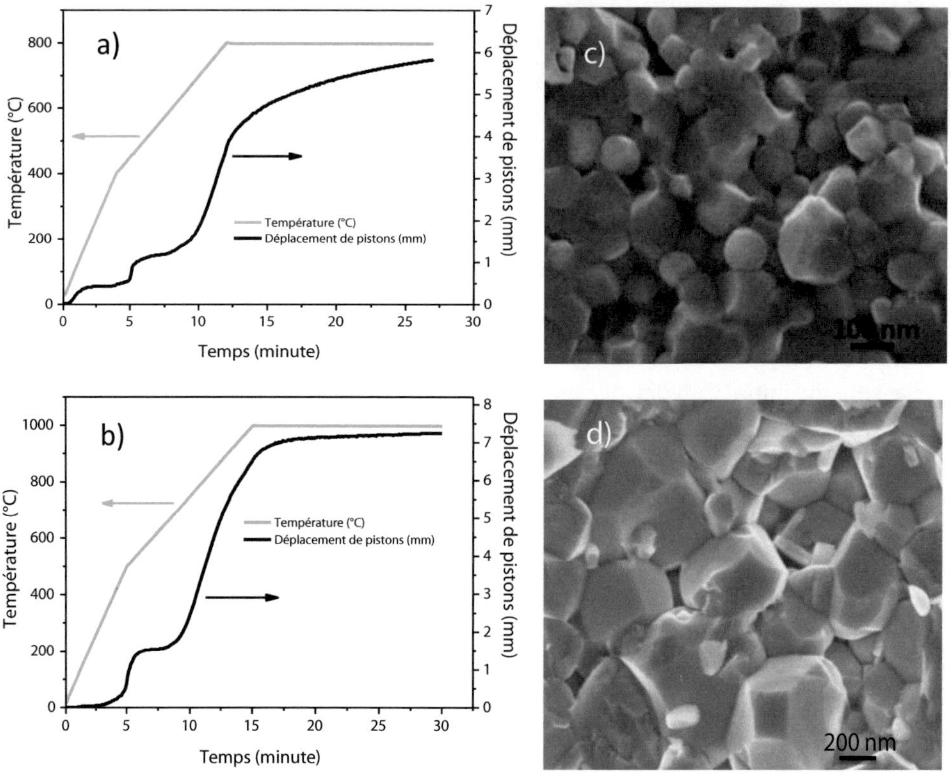

Figure 4. Shrinkage curve (temperature and piston displacement as a function of time) recorded during the RSPS process of $La_{0.85}Na_{0.15}MnO_3$ (a) and $LaMnO_3$ (b) ceramics and representative SEM micrographs of each ceramic (c and d, respectively) (adapted from Ref. [42]).

Starting from a mixture of raw oxides assisted by ball milling has been extensively used for the mechanical activation of the reactive powders before SPS treatment [36–38]. In order to achieve high-density ceramics with low SPS temperatures, a small particle size and high reactivity must be taken into consideration [39]. As an example, $Ni_{0.5}Zn_{0.5}Fe_2O_4$ ferrite was prepared by Song et al. [40]. Stoichiometric quantities of NiO, ZnO, and Fe_2O_3were milled in a high-energy planetary ball mill. Different parameters of the grinding time were varied, for example 10, 20, and 40 h, at a speed of 400 rpm. As expected, they found the higher the grinding time, the smaller the average size for the starting powders (<100 nm). For the sintering process, they selected the powders with 40 h milling time and particle size below 100 nm. Different temperatures were chosen (850, 875, 900, and 925°C) for 5 min, a pressure of 48 MPa and 5-min vacuum were applied. The best densification result was obtained at 925°C. A density of 5.23 g/cm^3, corresponding to 99% of the theoretical value, was reached. No secondary phases were detected in structural characterization. Zehani et al. [41] studied NiZnCu ferrite at several grinding times and speeds. The stoichiometric proportions of precursor oxides (NiO, Fe_2O_3, CuO, ZnO) were ground in a planetary mill. RSPS was then performed at different temperatures and holding times, using a graphite die and working under argon atmosphere. The main conclusion was that the final particle size increased with increasing milling speed. Also, in milling times shorter than 2 h at 800 rpm, the lattice parameter variation was insignificant.

Amorphous or poorly crystallized intermediate solid phases obtained by soft chemistry, combining precipitation in a liquid solution and moderate annealing, and containing all the desired elements were also used to form highly dense and fine-grained oxide ceramics. This is for instance the case of $La_{0.65}Ca_{0.20}Na_{0.15}MnO_3$ manganite [43] and $Y_3Fe_5O_{12}$ garnet [49] ceramics. The precipitated solids were first annealed at 600 and 400°C, respectively, to remove the main noninorganic species in the form of H_2O and CO_2, and then SPS treated at a temperature of 700 and 750°C, respectively, to obtain highly dense and fine-grained ceramics.

5. RSPS-made magnetic ceramics: Synthesis and properties

5.1. Soft magnets: Garnets

Magnetic garnets possess the crystal structure of mineral $Mn_3Al_2Si_3O_{12}$, with rare-earth (RE) and Fe^{3+} cations instead, leading to the general formula $RE_3Fe_5O_{12}$, RE is in the series from $La^{3+}$to Lu^{3+}. One of the most studied phases is the yttrium iron garnet (YIG), which is a remarkable ferrimagnetic material with many applications in microwave [44], magnetooptical [45], and spintronic devices [46], most of them based on the fact that YIG has the smallest linewidth for ferromagnetic resonance (FMR) [44]. Its ferrimagnetism results from

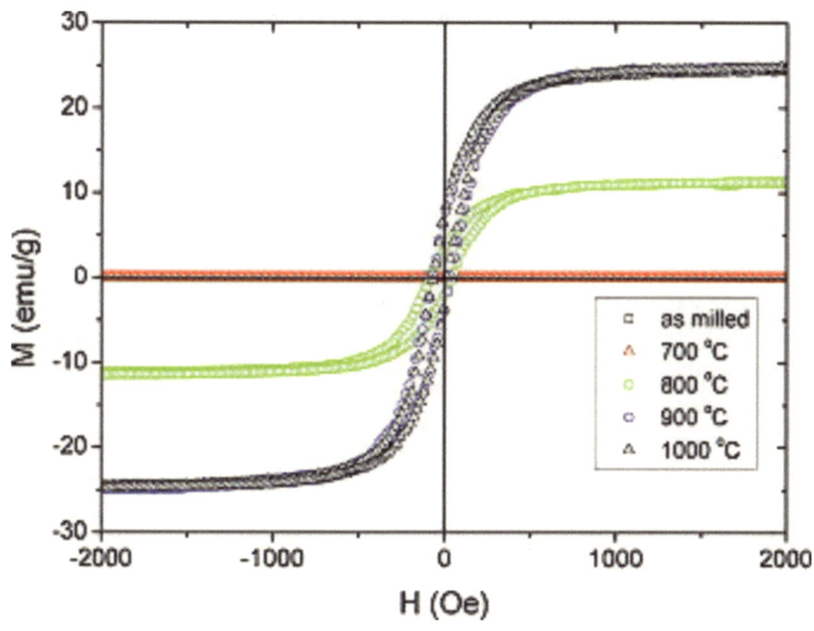

Figure 5. Hysteresis loops of ball-milled mixtures of iron and yttrium oxide for 5 h and annealed for 3 h at different temperatures (adapted from Ref. [48]).

superexchange interactions [55] between octahedral and tetrahedral Fe^{3+} cations, which are antiparallel. As a bulk, YIG is commonly prepared by the classic solid-state reaction technique which involves temperatures as high as 1350°C, for a few hours [47].

Nanostructured YIG (or other garnet) is typically prepared by combining soft chemistry, or ball milling, and annealing to complete the reaction before SPS sintering. This garnet crystal structure possesses a relatively large unit cell (160 atoms), making it difficult to achieve its synthesis at low temperature. A typical procedure can be high-energy ball milling of Fe_2O_3 + Y_2O_3oxide reagents, followed by thermal annealing before SPS treatment [48]. The magnetization of the recovered annealed powders increases with the increase in the annealing temperature, and it approaches its bulk value, namely 28 emu/g, only by samples annealed at $T \geq 900$°C (**Figure 5**).

RSPS offers an excellent alternative to produce consolidated nanostructured YIG at low temperatures and very short sintering times. A convenient YIG precursor can be prepared by hydrolysis in a polyol method, followed by processing by RSPS to obtain a nanostructured garnet phase [49], with the general magnetic properties of bulk YIG. The intermediate solid phase is amorphous (**Figure 6**) with the required Y/Fe stoichiometric ratio. Its preannealing at a temperature of 400°C allows its decomposition and the removal of main organic contents, but it is unsuitable to form the desired garnet phase. A reaction/sintering RSPS treatment at 750°C for 15 min produced a nanostructured solid with high density and nanosize grains. XRD-resolved patterns, **Figure 6**, showed that an amorphous phase leads first to the orthoferrite $YFeO_3$ phase (600–650°C) and then the transformation to the garnet $Y_3Fe_5O_{12}$phasefrom 750°C.

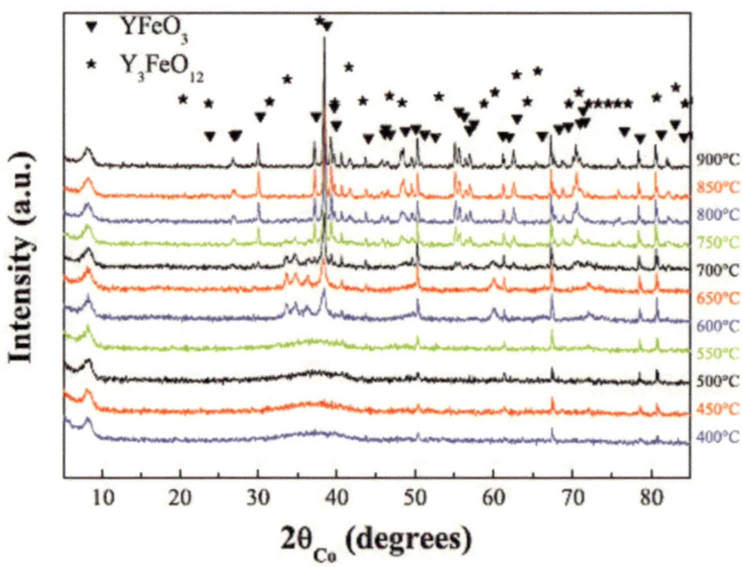

Figure 6. Temperature-resolved X-ray diffraction patterns of the polyol-synthesized YIG precursor. At about 600°C, yttrium orthoferrite ($YFeO_3$) is formed, which then transforms into YIG at higher temperatures [49].

Interestingly, the resulting dense and submicrometer-grain-sized ceramic exhibited the same magnetic properties as the conventionally made bulk counterpart: a saturation magnetization of 28 emu/g and a coercive field close to zero at room temperature. Clearly, the reduction of the grain size from the micrometer size range to the submicrometer one does not introduce major magnetic changes, the surface-to-volume atomic fraction remaining negligible in both cases to induce significant magnetic changes.

5.2. Magnetocalorics: Manganites

P-doped manganite $Ln_{1-x}X_xMnO_3$ (Ln: trivalent rare-earth ion; X: monovalent alkaline, or divalent alkaline-earth cation) phases with the perovskite structure have been extensively studied over the last 15 years in view of their remarkable physical properties, which can be used for a wide variety of applications, particularly for giant magnetoresistance devices and magnetocalorics [50–53]. The correlation between magnetic and transport properties are interpreted on the basis of double exchange (DE) mechanism [54], the superexchange (SE) interactions [55], the electron-phonon coupling due to the Jahn-Teller effect of Mn^{3+} ions, and the magnetic-phase separation [51]. The undoped stoichiometric $LaMnO_3$ compound, containing only Mn^{3+} ions is an insulating antiferromagnet [56], while doped ones contain Mn^{3+} and Mn^{4+} ions and may be ferromagnetic conductors. The magnetic properties of the former are driven by SE interactions, while those of the latter are mainly due to DE interactions. Consequently, the

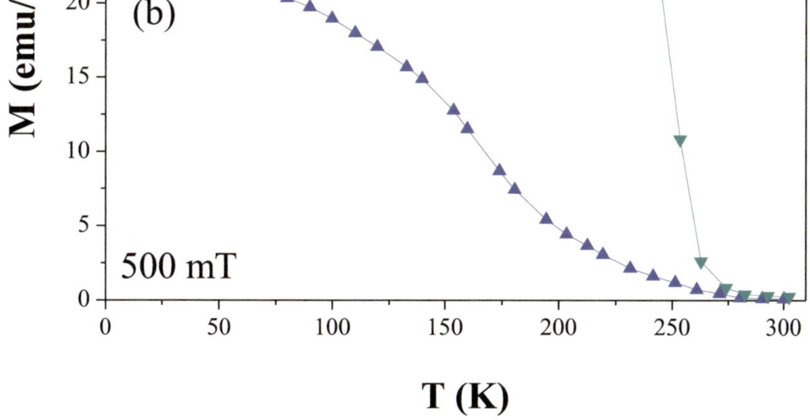

Figure 7. Temperature dependence of magnetization at 50 mT of $La_{0.70}Ca_{0.30}MnO_3$ ceramic produced by (a) conventional solid-state route at 1300°C and (b) by RSPSP at 800°C (adapted from Ref. [59]).

atomic Mn^{4+}/Mn^{3+} ratio is a key parameter in the achievement of the magnetoelectrical properties of these oxides.

To date, RSPS was successfully used to produce various manganite solid solutions, starting from raw oxides or hydroxides annealed at 800–1000°C under an uniaxial pressure of 50 MPa and under vacuum: $LaMnO_3$ [42], $La_{0,85}Na_{0,15}MnO_3$ [57], $La_{0,85}Na_{0,15-x}K_xMnO_3$ [12], $La_{0.67}Ca_{0.33}MnO_3$ [58], and $La_{0.7à}Ca_{0.30-x}Ba_xMnO_3$ [59] among others. Interestingly, all the produced ceramics exhibited high densities over 90% of the theoretical values and submicrometer grain size, and a systematically broadened paramagnetic-to-ferromagnetic transition as a function of temperature, with a decreased Curie temperature (T_C) value (**Figure 7**). These changes are due to an evolution of the chemical composition concerning the synthesis conditions. The conventionally made ceramics are assumed to be chemically homogeneous with a Mn^{4+}/Mn^{3+}atomic ratio fixed by the doping rate, while the RSPS-made ones may suffer from heterogeneities related to their very rapid reacting/sintering kinetics. These heterogeneities can be associated to a Mn^{4+} concentration variation between the ceramic core and its surface in contact with graphite during RSPS experiments, with a total Mn^{4+}/Mn^{3+} atomic ratio smaller than its theoretical value. This discrepancy was confirmed by K-Mn edge X-ray absorption spectroscopy (XANES) and iodometry chemical analysis and related to the reductive SPS processing conditions [58].

To evaluate the magnetocaloric properties of manganites, the variation of the magnetic entropy upon a given magnetic field change, ΔM, is usually inferred from the first magnetization curves and plotted as a function of the temperature around T_C value. ΔM (T) of RSPS-processed manganites is systematically much more broadened than that of their conventionally made ceramics (**Figure 8**) for the reasons given above. The commercial applications require a magnetocaloric effect extending on a broad temperature range. RSPS ceramics may just offer such an opportunity.

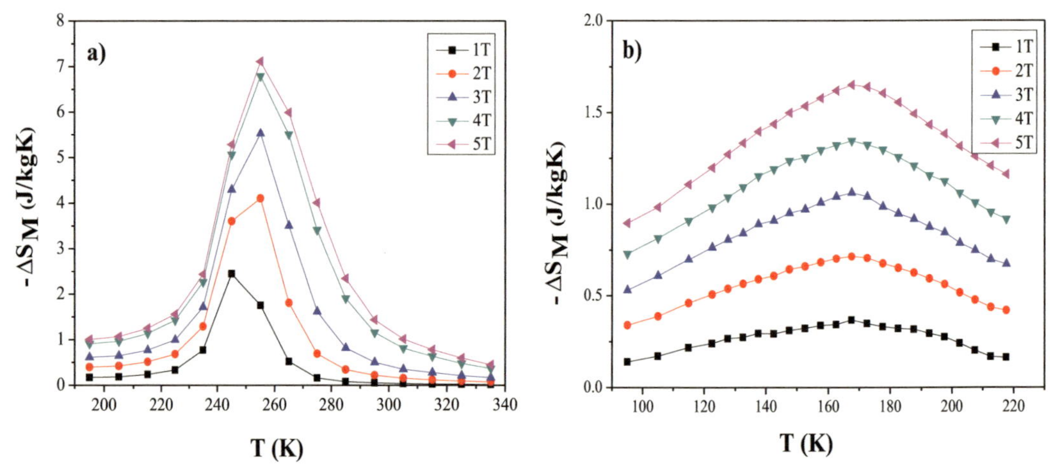

Figure 8. Magnetic entropy variation ΔS_M as a function of the temperature for a magnetic field change of 1, 2, 3, 4, and 5T $La_{0.70}Ca_{0.30}MnO_3$ ceramic produced by conventional solid-state route at 1300°C (a) and by RSPS at 800°C (adapted from Ref. [59]).

5.3. Hard magnets: Hexaferrites

Hexaferrites have become extremely important materials since they have a large variety of applications due to their high magnetocrystalline anisotropy in relation with their ΔM hexagonal structure. Their magnetic properties are mainly driven by SE interactions as most of insulating oxides. Technologically speaking, hexaferrites are mainly used in the form of bulk solids as permanent magnets in magnetic recording and magnetic data storage devices, and more recently, as systems operating at microwave/GHz frequencies [60].

As nanopowders, hexaferrites can be produced by different methods such as sol-gel [61], hydrothermal [62], aerosol pyrolysis [63], or mechanochemical synthesis [64]. In most of the cases, a subsequent annealing is required to provide enough energy to complete phase formation and an SPS sintering is needed to achieve their consolidation.

To the best of our knowledge, the RSPS process has been scarcely used to produce consolidated nanostructured hexaferrites. Bolarín-Miró et al. [65] carried out a comparative study between M-type strontium hexaferrite prepared from strontium and iron single oxides mechanically activated by high-energy ball milling for 5 h followed by RSPS, and the same milled powder mixture sintered by conventional route. They showed that, in comparison with conventional heat treatment, RSPS process allows the formation of strontium hexaferrite single phase at lower temperatures with a higher magnetization. In contrast, the resulting ceramics exhibited smaller coercive field (**Figure 9**).

Stingaciu et al. [66] reported the preparation of strontium hexaferrite by SPS starting also from ball-milling-activated commercially available $SrFe_{12}O_{19}$ powder. They observed that the hexaferrite phase is maintained during the milling process (8 to 42 h), while it is not during the SPS treatment. Due to the reductive operating conditions, a nonnegligible amount of magnetite is formed leading to the production of a $SrFe_{12}O_{19}$-Fe_3O_4 nanocomposite. Moreover, they evidenced a pronounced decrease in the room-temperature coercive field (H_c) and an increase

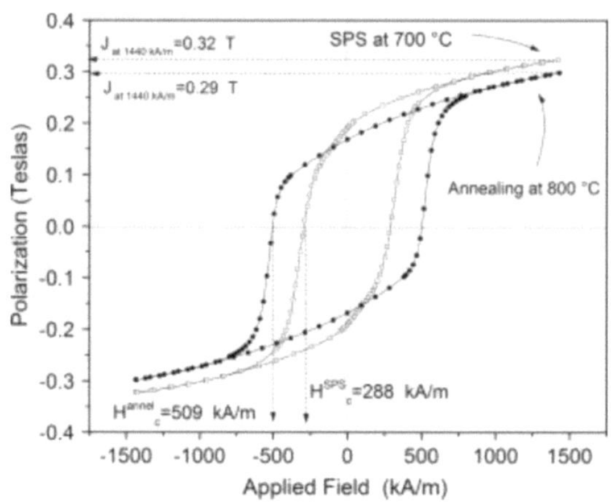

Figure 9. *J(H)* Hysteresis loops of precursor's mixtures milled for 5 h and (a) pressed at 800 MPa and annealed at 800°C (circle) and (b) SPS-treated at 700°C under 80 MPa (square) (adapted from Ref. [65]).

in the magnetization (*Ms*) at maximum applied field of 1 T, for powders milled for a longer time and consolidated (**Figure 10**). They concluded that the magnetic properties of the studied nanocomposites are largely conditioned by the extrinsic properties of the secondary phase,

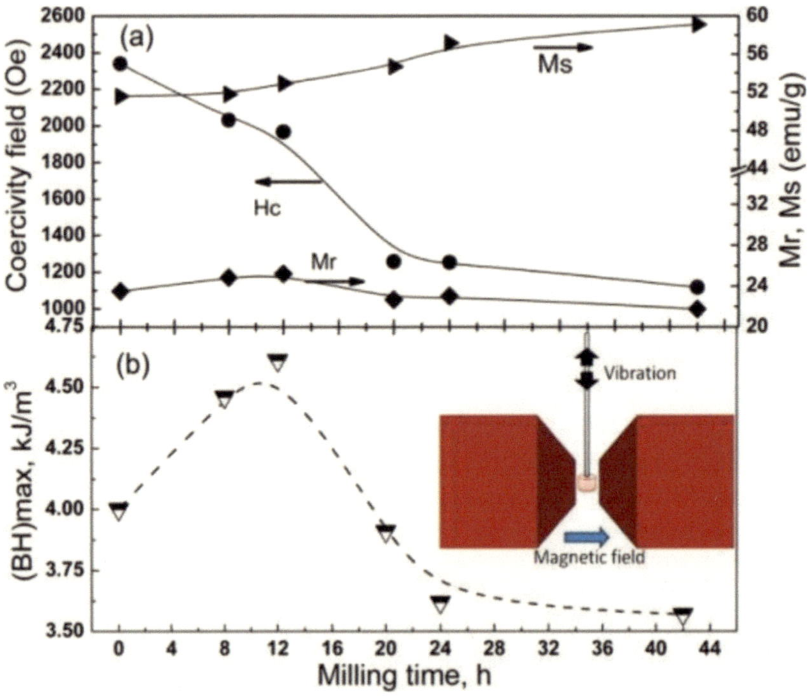

Figure 10. Evolution of the coercive field *Hc* (a), the remanent magnetization *Mr* (a), the magnetization at maximum applied field 1 T, *Ms*, and the maximum magnetic energy product $(BH)_{max}$ of SPS-consolidated $SrFe_{12}O_{19}$ powder after different milling times. These data obtained from the hysteresis loops recorded at room temperature, applying the external magnetic field perpendicular to the uniaxial SPS pressing direction [66].

Fe_3O_4, a soft ferrimagnet, formed after the SPS, rather than the hard $SrFe_{12}O_{19}$phaseparticle-size-reduction effect. Additionally, they reported the highest maximal magnetic energy values $(BH)_{max}$, 4.0–4.6 kJ/m^3, for the samples with the lowest Fe_3O_4 content, underlining the complexity of the involved demagnetization mechanism.

Vázquez-Victorio [67] combined soft chemistry synthesis (polyol process) and consolidation by SPS to produce nanostructured $BaFe_{12}O_{19}$ barium hexaferrite. Typically, they produced an intermediate solid phase by reaction of the metallic salts in a polyol within an appropriate Ba/Fe atomic ratio; they were annealed at 800°C to complete the desired crystalline phase before SPS sintering at 800°C for 5–10 min and 100 MPa, under vacuum. Varying the nature of the metallic salts and the polyol solvent, they succeeded to produce highly dense (density > 95%) and ultrafine-grained (~100 nm) pure $BaFe_{12}O_{19}$and $BaFe_{12}O_{19}$ with a small content of iron oxide. A direct dependency of the magnetic properties of the produced solids on their iron oxide content was observed (**Figure 11**). The highest coercive field and magnetization

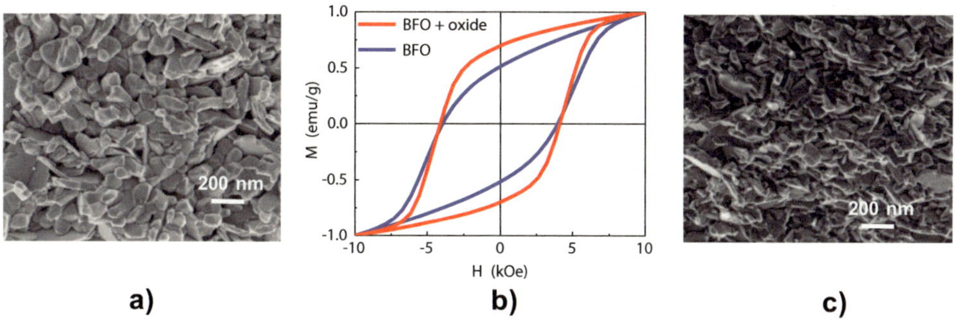

Figure 11. SEM micrographs of (a) $BaFe_{12}O_{19}$ and (c) $BaFe_{12}O_{19}$-Fe_2O_3 nanostructures produced by combining polyol process to SPS (800°C, 100 MPa, 5 min), and their room temperature hysteresis loops (b) [67].

at maximum applied field of 1 T (and hence the energy product *BH*) was measured at room temperature, on the nanocomposite. The hysteresis loops also appeared closer to a rectangular form, which is the best shape for applications, as this leads to a well-defined coercive field and remanent magnetization.

6. Conclusions

Currently, spark plasma sintering appears as the only method capable to consolidate nanopowders into high-density nanostructured solids; in this chapter, we have briefly reviewed its application to carry out also the solid-state reaction needed to achieve a particular phase, starting from precursors synthesized by diverse methods. Many challenges remain, especially in the cases of reaction by precursor decomposition, when reaction and sintering temperatures are significantly different. RSPS is still a very young technique, with many potential capabilities, which will certainly be developed in the near future.

Author details

Giulia Franceschin[1], Nancy Flores-Martínez[1], Gabriela Vázquez-Victorio[1,2], Souad Ammar[1] and Raul Valenzuela[1,2]*

*Address all correspondence to: monjaras@unam.mx

1 ITODYS, Université Paris Diderot, Sorbonne Paris Cité, Paris Cedex, France

2 Institute for Materials Research, National Autonomous University of Mexico, Mexico City, México

References

[1] Anis SF, Khalil A, Saepurahman, Singarave G, Hashaikeh R. A review on the fabrication of zeolite and mesoporous inorganic nanofibers formation for catalytic applications. Microporous & Mesoporous Materials. 2016;**236**:176-192. DOI: 10.1016/j.micromeso.2016.08.043

[2] Zeinali-Sehrig F, Majidi S, Asvadi S, Hsanzadeh A, Hossein-Rasta S, Emamverdy M, Akbarzadeh J, Jahangiri S, Farahkhiz S, Akbarzadeh A. An update on clinical applications of magnetic nanoparticles for increasing the resolution of magnetic resonance imaging. Artificial Cells, Nanomedicine and Biotechnology. 2016;**44**:1583-1588. DOI: 10.3109/21691401.2015.1101001

[3] Kuhl S, Strasser P. Oxygen electrocatalysis on dealloyed Pt nanocatalysts. Topics in Catalysis. 2016;**59**:1628-1637. DOI: 10.1007/s11244-016-0682-z

[4] Xiao D, Lu T, Zeng R, Bi Y. Preparation and highlighted applications of magnetic microparticles and nanoparticles: A review on recent advances. Microchimica Acta. 2016;**183**:2655-2675. DOI: 10.1007/s00604-016-1928-y

[5] Guimaraes AP. Principles of Nanomagnetism. Berlin, Heidelberg: Springer-Verlag; 2009. p. 236. DOI: 10.1007/978-3-642-01482-6

[6] Bedanta S, Barman A, Kleemann W, Petracic O, Seki T. Synthesis, properties, and applications of single-domain magnetic nanoparticles. Journal of Nanomaterials. 2013:130180. DOI: 10.1155/2013/130180

[7] Thomson T, Hu G, Terris BD. Intrinsic distribution of magnetic anisotropy in thin films probed by patterned nanostructures. Physical Review Letters. 2006;**96**:257204. DOI: 10.1103/PhysRevLett.96.257204

[8] Orrù R, Licheri R, Locci AM, Cincotti A, Cao G. Consolidation/synthesis of materials by electric current activated/assisted sintering. Materials Science and Engineering Review. 2009;**63**:127-287. DOI: 10.1016/j.mser.2008.09.003

[9] Guillon O, Gonzalez-Julian J, Dargatz B, Kessel T, Schierning G, Räthel J, Herrmann M. Field-assisted sintering technology/spark plasma sintering: Mechanisms, materials, and technology developments. Advanced Engineering Materials. 2014;**16**:830-849. DOI: 10.1002/adem.201300409

[10] Garay JE. Current-activated, pressure-assisted densification of materials. Annual Reviews of Materials Science. 2010;**40**:445-468. DOI: 10.1146/annurev-matsci-07909-104433

[11] Kali R, Mukhopadhyay A. Spark plasma sintered/synthesized dense and nanostructured materials for solid-state Li-ion batteries: Overview and perspective. Journal of Power Sources. 2014;**920**:920-931. DOI: 10.1016/j.jpowsour.2013.09.010

[12] Regaieg Y, Sicard L, Monnier J, Koubaa M, Ammar-Merah S, Cheikhrouhou A. Magnetic and magnetocaloric properties of La0.85(Na1−xKx)0.15MnO3 ceramics produced by

reactive spark plasma sintering. Journal of Applied Physics. 2014;**115**:17A917. DOI: 10.1063/1.4864138

[13] Navrotsky A. Energetics of oxide nanoparticles. International Journal of Quantum Chemistry. 2009;**109**:2647. DOI: 10.1002/qua.21981

[14] Production of sintering components. In: Höganäs Handbook for Sintering Components. 2013. pp. 25-49, Höganäs Ab Laboratories, Sweden. Website: Handbook2_Production_of_Sintered_Components_December_2013_0675HOG_interactive.pdf

[15] Isobe T, Daimond K, Sato T, Matsubara T, Hikichia Y, Ota T. Spark plasma sintering technique for reaction sintering of Al_2O_3/Ni nanocomposites and its mechanical properties. Ceramics International. 2008;**34**:213-217. DOI: 10.1016/j.ceramint.2006.08.017

[16] Available from: http://web.mit.edu/2.813/www/readings/Ellingham_diagrams.pdf

[17] Liu H, Liu L, Ye F, Zhang Z, Yu Z. Microstructure and mechanical properties of the spark plasma sintered TaC/SiC composites: Effects of sintering temperatures. Journal of European Ceramics Society. 2012;**32**:3617-3825. DOI: 10.1016/j.jeurceramsoc.2012.04.042

[18] Monteverde F. Ultra-high temperature HfB_2–SiC ceramics consolidated by hot-pressing and spark plasma sintering. Journal of Alloys and Compounds. 2007;**428**:197-205. DOI: 10.1016/j.jallcom.2006.01.107

[19] Sciti D, Bonnefont G, Fantozzi G, Silvestroni L. Spark plasma sintering of HfB_2 with low additions of silicides of molybdenum and tantalum. Journal of the European Ceramic Society. 2010;**30**:3253-3258. DOI: 10.1016/j.jeurceramsoc.2010.06.006

[20] Licheri R, Orrù R, Musa C, Locci AM, Cao G. Consolidation via spark plasma sintering of HfB_2/SiC and HfB_2/HfC/SiC composite powders obtained by self-propagating high-temperature synthesis. Journal of Alloys and Compounds. 2009;**478**:572-578. DOI: 10.1016/j.jallcom.2008.11.092

[21] Casolco SR, Xu J, Garay JE. Transparent/translucent polycrystalline nanostructured yttria stabilized zirconia with varying colors. Scripta Materalia. 2008;**58**:516-519. DOI: 10.1016/j.scriptamat.2007.11.014

[22] Ayadi F, Ammar S, Cheikhrouhou-Koubaa W, Cheikhrouhou A, Nowak S, Monnier J, Sicard. Magnetocaloric nanostructured $La_{0.7}Ca_{0.3-x}Ba_xMnO_3$ ($0<x<0.3$) ceramics produced by combining polyol process to spark plasma sintering. Journal of Alloys and Compounds. 2017;**691**:474-481. DOI: 10.1016/j.jallcom.2016.08.281

[23] Valenzuela R, Gaudisson T, Ammar S. Severe reduction of Ni-Zn ferrites during consolidation by spark plasma sintering (SPS). Journal of Magnetism and Magnetic Materials. 2016;**400**:311-314. DOI: 10.1016/j.jmmm.2015.07.044

[24] Grasso S, Yoshida H, Porwala H, Sacca Y, Reece M. Highly transparent α-alumina obtained by low cost high pressure SPS. Ceramics International. 2013;**39**:3243-3248. DOI: 10.1016/j.ceramint.2012.10.012

[25] Weissler GA. Resistance sintering with alumina dies. International Journal of Powder Metallurgy. 1981;**17**:107

[26] Kamikawa M, Kano Y. JP Patent JP2001226703; 2001

[27] Munir AZ, Anselmi-Tamburini U, Ohyanagi M. The effect of electric field and pressure on the synthesis and consolidation of materials: A review of the spark plasma sintering method. Journal of Materials Science. 2006;**41**:763-777. DOI: 10.1007/s10853-006-6552-2

[28] Dudina DV, Mukherjee AK. Reactive spark plasma sintering: Successes and challenges of nanomaterial synthesis. Journal of Nanomaterials. 2013:625218. DOI: 10.1155/2013/625218

[29] Schmidt J, Niewa R, Schmidt M, Grin Y. Spark plasma sintering effect on the decomposition of MgH_2. Journal of the American Ceramic Society. 2005;**88**:1870-1874. DOI: 10.1111/j.1551-2916.2005.00358.x

[30] Skandan G, Hahn H, Kear BH, Roddy M, Cannon WR. The effect of applied stress on densification of nanostructured zirconia during sinter-forging. Materials Letters. 1994;**20**:305. SSDI: 0167-577x(94)00114-3

[31] Chen XJ, Khor KA, Chan SH, Yu LG. Overcoming the effect of contaminant in solid oxide fuel (SOFC) electrolyte: Spark plasma sintering (SPS) of 0.5 wt% silica-doped yttria-stabilized zirconia (YSZ). Materials Science and Engineering A. 2004;**374**:64-71. DOI: 10.1016/j.msea.2003.12.028

[32] Nygren M. SPS processing of nano-structured ceramics. Journal of Iron and Steel Research, International. 2007;**14**:99-103. DOI: 10.1016/S1006-706X(08)60060-6

[33] Diouf S, Molinari A. Densification mechanisms in spark plasma sintering: Effect of particle size and pressure. Powder Technology. 2012;**221**:220-227. DOI: 10.1016/j.powtec.2012.01.005

[34] Stanciu L, Groza JR, Stoica L, Plapcianu C. Influence of powder precursors on reaction sintering of Al_2TiO_5. Scripta Materialia. 2004;**50**:1259-1262. DOI: 10.1016/j.scriptamat.2004.01.034

[35] Gaudisson T, Artus M, Acevedo U, Herbst F, Nowak S, Valenzuela R, Ammar S. On the microstructural and magnetic properties of fine-grained $CoFe_2O_4$ ceramics produced by combining polyol process and spark plasma sintering. Journal of Magnetism and Magnetic Materials. 2014;**370**:87-95. DOI: 10.1016/j.jmmm.2014.06.014

[36] Munir ZA. Synthesis and densification of nanomaterials by mechanical and field activation. Journal of Materials Synthesis and Processing. 2000;**8**:189-196. DOI: 10.1023/A:1011312126285

[37] Zhang HW, Gopalan R, Mukai T, Hono K. Corrigendum to "Fabrication of bulk nanocrystalline Fe-C alloys by spark plasma sintering of mechanically milled powder" [Scripta Materialia 2005; 53: 863-868]. Scripta Materialia. 2005;**54**:1827-1828. DOI: 10.1016/j.scriptamat.2006.02.001

[38] Locci AM, Orrù R, Cao G, Munir ZA. Effect of ball milling on simultaneous spark plasma synthesis and densification of TiC–TiB_2 composites. Ceramics International. 2006;**434**:23-29. DOI: 10.1016/j.msea.2006.06.131

[39] Cullity BD. Introduction to Magnetic Materials. California: Addison-Wesley Publishing Company; 1972

[40] Song S, Song Q, Li J, Mudinepalli VR, Zhang Z. Characterization of submicrometer-sized NiZn ferrite prepared by spark plasma sintering. Ceramics International. 2014;**40**:6473-6479. DOI: 10.1016/j.ceramint.2013.11.099

[41] Zehani K, Ahmadi B, Loyau V, Champion Y, Bessais L, Lobue M, Labouré E, Mazaleyrat F. Structural, dielectric, and magnetic properties of NiZnCu ferrites synthesized by reactive spark plasma sintering process. IEEE Transactions on Magnetics. 2014;**50**:1-4. DOI: 0.1109/TMAG.2013.2284766

[42] Regaieg Y, Delazair G, Herbst F, Sicard L, Monnier J, Montero D, Villeroy B, Koubaa M, Ammar S, Cheikhrouhou A, Godart C. Fast reactivity by spark plasma sintering: The case of orthorhombic $LaMnO_3$ compound. Material Letters. 2012;**80**:195-198. DOI: 10.1016/j.matlet.2012.04.046

[43] Abdallah Ben Ammar A, Ayadi F, Nowak S, Lecoq H, Cheikhrouhou-Koubaa W, Cheikhrouhou A, Ammar S, Sicard L. A combined sol–gel and spark plasma sintering route to produce highly dense and fine-grained $La_{0.65}Ca_{0.20}Na_{0.15}MnO_3$ ceramics for magnetocaloric applications. Materials Research Express. 2014;**1**:015703. DOI: 10.1088/2053-1591/1/1/015703

[44] Mo N, Green JJ, Beitscher BA, Patton CE. High precision-metrology based microwave effective line width measurement technique. Review of Scientific Instruments. 2007;**78**:113903. DOI: 10.1063/1.2813344

[45] Stadler BJH, Mizumoto T. Integrated magneto-optical materials and isolators. IEEE Photonics Journal. 2013;**6**:0600215. DOI: 10.1109/JPHOT.2013.2293618

[46] Hyde P, Bai L, Kumar DMJ, Southern BW, Hu C-M, Huang SY, Miao BF, Chien CL. Electrical detection of direct and alternating spin current injected from a ferromagnetic insulator into a ferromagnetic metal. Physical Review B. 2014;**89**:180404. DOI: 10.1103/PhysRevB.89.180404

[47] Niyaifar Ramani M, Radhakrishna MC, Hasapour A, Mozaffari M, Amighian J. Magnetic studies of $Bi_xY_{3-x}Fe_5O_{12}$ fabricated using conventional method. Hyperfine Interactions. 2008;**187**:137-141. DOI: 10.1007/s10751-008-9875-6

[48] Sánchez-De Jesús F, Cortés CA, Valenzuela R, Ammar S, Bolarín-Miró AM. Synthesis of $Y_3Fe_5O_{12}$ (YIG) assisted by high-energy ball milling. Ceramics International. 2012;**38**:5257-5263. DOI: 10.1016/j.ceramint.2012.03.036

[49] Gaudisson T, Acevedo U, Nowak S, Yaacoub N, Greneche J-M, Ammar S, Valenzuela R. Combining soft chemistry and spark plasma sintering to produce highly dense and finely grained soft ferrimagnetic $Y_3Fe_5O_{12}$ (YIG) ceramics. Journal of the American Ceramic Society. 2013;**96**:3094-3099. DOI: 10.1111/jace.12452

[50] Jin S, Tiefel TH, McCormack M, Fastnacht RA, Ramesh R, Chen LH. Thousand fold change in resistivity in magnetoresistive La-Ca-Mn-O films. Science. 1994;**264**:413-415. DOI: 10.1126/science.264.5157.413

[51] Yunoki S, Hu J, Malvezzi AL, Moreo A, Furukawa N, Dagotto E. Phase separation in electronic models for manganites. Physical Review Letters. 1998;**80**:845. DOI: 10.1103/PhysRevLett.80.845

[52] Salamon MB, Jaime M. The physics of manganites: Structure and transport. Reviews of Modern Physics. 2001;**73**:583. DOI: 10.1103/RevModPhys.73.583

[53] Phan M-H, Yu S-C. Review of the magnetocaloric effect in manganite materials. Journal of Magnetism and Magnetic Materials. 2007;**308**:325-340. DOI: 10.1016/j.jmmm.2006.07.025

[54] Zener N. Interaction between the d-shells in the transition metals. II. Ferromagnetic compounds of manganese with perovskite structure. Physical Review. 1951;**82**:403. DOI: /10.1103/PhysRev.82.403

[55] Goodenough JB. Theory of the role of covalence in the perovskite-type manganites [La,M(II)]MnO_3. Physical Review. 1955;**100**:564. DOI: 10.1103/PhysRev.100.564

[56] Topfer J, Goodenough JB. Journal of Solid State Chemistry. $LaMnO_{3-\delta}$ Revisited. 1997;**130**:117-128. DOI: 10.1006/jssc.1997.7287

[57] Regaieg Y, Sicard L, Monnier J, Delaizir G, Koubaa M, Godart C, Ammar-Merah S, Cheikhrouhou A. Rapid synthesis of $La_{0.85}Na_{0.15}MnO_3$ by spark plasma sintering: Magnetic and magnetocaloric study. Materials Chemistry and Physics. 2013;**139**:629-633. DOI: 10.1016/j.matchemphys.2013.02.008

[58] Regaieg Y, Ayadi F, Nowak S, Koubaa M, Cheikhrouhou A, Reguer S, Monnier J, Sicard L, Ammar-Merah S. Magnetocaloric properties of $La_{0.67}Ca_{0.33}MnO_3$ produced by reactive spark plasma sintering and by conventional ceramic route. Material Research Express. 2014;**1**:046105. DOI: 10.1088/2053-1591/1/4/046105

[59] Ayadi F. Approche couplée de synthèse par chimie douce et SPS de manganites substitués pour la réfrigération magnétique [joint PhD thesis]. France and Tunisia: Paris Diderot University and University of Sfax; 2014

[60] Pullar RC. Hexagonal ferrites: A review of the synthesis, properties and applications of hexaferrite ceramics. Progress in Materials Science. 2012;**57**:1191-1334. DOI: 10.1016/j.pmatsci.2012.04.001

[61] Jamalian M. An investigation of structural, magnetic and microwave properties of strontium hexaferrite nanoparticles prepared by a sol–gel process with doping SN and Tb. Journal of Magnetism and Magnetic Materials. 2015;**378**:217-220. DOI: 10.1016/j.jmmm.2014.11.047

[62] Drofenik M, Ban I, Makovec D, Žnidaršič A, Jagličić Z, Hanžel D, Lisjak D. The hydrothermal synthesis of super-paramagnetic barium hexaferrite particles. Materials Chemistry and Physics. 2011;**127**:415-419. DOI: 10.1016/j.matchemphys.2011.02.037

[63] González-Carreño T, Morales MP, Serna CJ. Barium ferrite nanoparticles prepared directly by aerosol pyrolysis. Materials Letters. 2000;**43**:97-101. DOI: 10.1016/S0167-577X(99)00238-4

[64] Sánchez-De Jesús F, Bolarín-Miró AM, Cortés-Escobedo CA, Valenzuela R, Ammar S. Mechanosynthesis, crystal structure and magnetic characterization of M-type $SrFe_{12}O_{19}$. Ceramics International. 2014;**40**:4033-4038. DOI: 10.1016/j.ceramint.2013.08.056

[65] Bolarín-Miró AM, Sánchez-De Jesús F, Cortés-Escobedo CA, Díaz-De la Torre S, Valenzuela R. Synthesis of M-type $SrFe_{12}O_{19}$ by mechanosynthesis assisted by spark plasma sintering. Journal of Alloys and Compounds. 2015;**643**:S226-S230. DOI: 10.1016/j.jallcom.2014.11.124

[66] Stingaciu M, Topole M, McGuiness P, Christensen M. Magnetic properties of ball-milled $SrFe_{12}O_{19}$ particles consolidated by spark-plasma sintering. Scientific Reports. 2015;**5**:14112. DOI: 10.1038/srep14112

[67] Vázquez-Victorio G. Ferritas nanoestructuradas, síntesis y propiedades magnéticas [PhD thesis]. Universidad Nacional Autónoma de Mexico (UNAM); 2017

Chapter 7

Selective Laser Sintering of Nanoparticles

Sukjoon Hong

Additional information is available at the end of the chapter

http://dx.doi.org/10.5772/intechopen.68872

Abstract

Selective laser sintering of nanoparticles has received much attention recently as it enables rapid fabrication of functional layers including metal conductors and metal-oxide electrodes on heat-sensitive polymer substrate in ambient conditions. Photothermal reactions induced by lasers rapidly increase the local temperature of the target nanoparticle in a highly selective manner, and subsequent sintering steps including melting and coalescence between nanoparticles occur to fabricate interconnected sintered films for various future applications. The mechanism of laser sintering, as well as possible target materials subject to laser sintering, together with experimental schemes developed to improve the process and potential applications, is briefly summarized in this chapter.

Keywords: laser, optics, nanoparticle, metal, metal-oxide, flexible electronics

1. Introduction

In this chapter, we focus on a specific type of sintering that utilizes laser and nanoparticle as its heat source and target material, respectively. Laser and nanoparticle on their own have interesting properties which are advantageous for conventional sintering process. Laser is a tool with a broad range of parameters and enables numerous responses such as remote temperature manipulation and rapid processing speed which cannot be achieved by other mechanical tools. Nanoparticles, having controllable sizes and shapes, find their application in various fields, and melting temperature depression due to their size effect is one of the key properties for sintering as it reduces the temperature required for the sintering process to a great extent. Laser sintering of nanoparticles—combining these two elements—not only possesses both the abovementioned features but also provides additional virtues and allows facile, damage-free fabrication of functional layers on heat-sensitive substrate to bring novel applications in the form of flexible electronics. Selective laser sintering of nanoparticles is summarized in this chapter with special emphasis on its mechanism, target material, experimental schemes, and potential applications.

2. Mechanism

Sintering of nanoparticles with other heat sources such as furnace or convection oven has been long studied. In a conventional process, nanoparticles in the form of ink are coated on a substrate with wet processing or diverse printing techniques, followed by bulk heating of the substrate at an elevated temperature. Thermal sintering usually involves long sintering time (>30 min) at high temperatures (>200°C) which are often not compatible with numerous heat-sensitive polymer substrates [1]. The main difference of selective laser sintering is that the heat source is substituted with laser. As a laser beam is focused on the designated spot, the optical energy is directly converted into heat through photothermal reaction to control the local temperature with high selectivity and controllability. Laser processing, including selective laser sintering of nanoparticles summarized in this chapter, is expected to show constant processing characteristics as far as the laser beam is properly controlled. Laser processing has certain advantages in terms of reproducibility compared to other processing techniques. Laser is an essentially massless and sterile tool so that the problems from mechanical holders or the processing tool itself can be largely prevented. Laser beam is not subject to wear and tear as well. These properties of laser help avoiding any contamination of the material being processed to increase the reproducibility of the proposed process.

Photothermal reactions by laser can be activated through various elementary excitation including interband and intraband transitions, and the detailed reaction depends on the type of absorbing material as well as the laser parameters [2]. We consider a continuous laser as a simple heat source as far as the characteristic time for the initial processing step is considerably longer than the relaxation time and no phase change occurs during the heating process. For pulsed laser, pulse duration and shape have strong effect on heating characteristics. A single pulse generally brings rapid temperature increase and subsequent cooling. As a result, the temperature swings up and down in case of multiple pulse irradiation, and the average temperature rise is dependent on the repetition rate [3].

Thermal properties of the target nanoparticle are important subjects in sintering, and it is still true for the laser sintering. Prior to the application of laser sintering, thermal properties of nanoparticle are often investigated in separate steps. The melting temperature of nanoparticle changes abruptly in general due to greatly enhanced surface-to-volume ratio at a nanoscale. Previous investigation on the thermal properties of silver nanoparticles is presented in Section 2 as a representative example. The melting point depression of silver nanoparticles at various sizes is calculated from the Gibbs-Thomson equation as shown in **Figure 1(a)** [4]. It is anticipated from the graph that the silver nanoparticle at ~5-nm diameter exhibits significantly lower melting temperature (~150°C) compared to its bulk counterpart at 960°C. A similar trend can be found in other nanoparticles as well [5–7]. Direct observation of the nanoparticle melting is possible through high-temperature tunneling electron microscope

http://dx.doi.org/10.5772/intechopen.68872

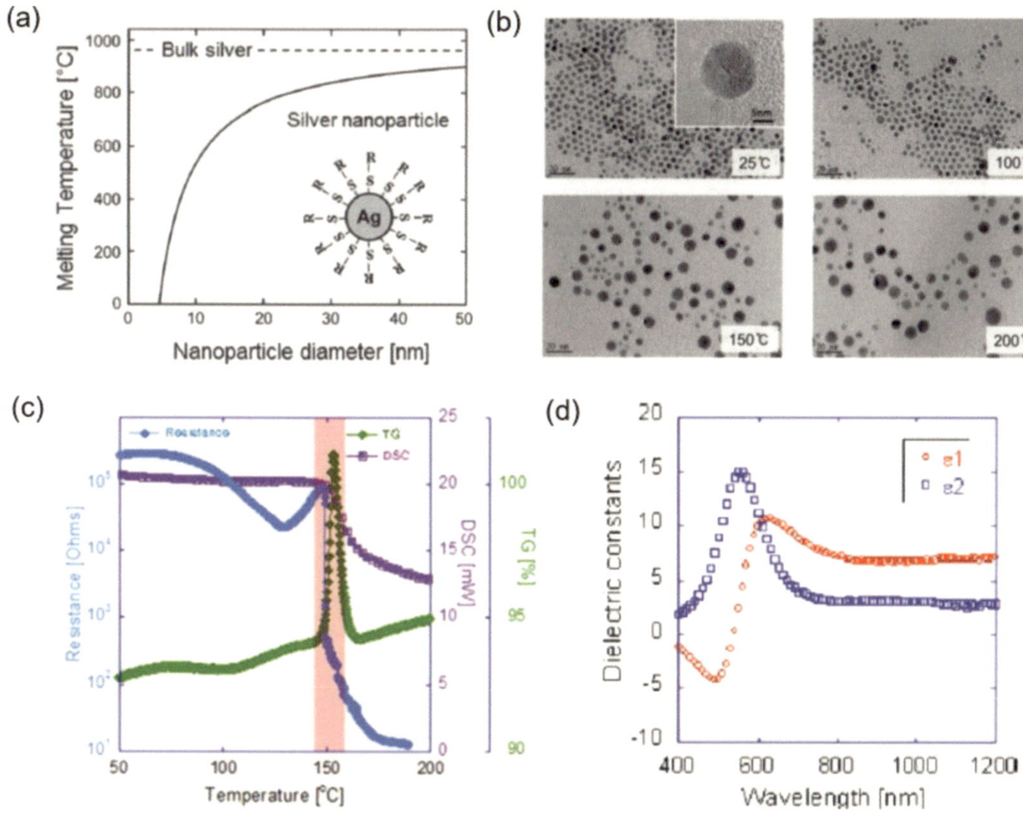

Figure 1. (a) The melting point depression of silver nanoparticles calculated from the Gibbs-Thomson equation. Reprinted with permission from Ref. [4]. Copyrights 2011 Wiley-VCH Verlag GmbH & Co. KGaA, Weinheim. (b) Observation of sintering behavior of silver nanoparticles at various temperatures by in-situ probing. Inset: HRTEM image of a silver nanoparticle with self-assembled monolayer (SAM). Reprinted with permission from Ref. [8]. Copyrights 2012 Elsevier Inc. (c) Melting characteristics of the silver nanoparticles in terms of resistance change (blue), TGA (green), and DSC (purple). Reprinted with permission from Ref. [9]. (d) Ellipsometric measurement of the silver nanoparticle film. Reprinted with permission from Ref. [11]. Copyrights 2008 American Institute of Physics.

(TEM) measurement as in **Figure 1(b)** [8]. The silver nanoparticles show dramatic increase in size when heated at 150°C, while there is no significant change up to 100°C, indicating that the significant melting and coalescence between nanoparticles occur at a temperature as low as 150°C. A more quantitative approach can be found from thermo-gravimetric analysis (TGA), differential scanning calorimetry (DSC), and electrical resistance measurement according to the temperature as shown in **Figure 1(c)** [9]. A sharp exothermic peak in DSC and a decrease in TGA at ~150°C indicate that there exists a phase change together with the decomposition of the capping organic molecules at the corresponding temperature. The electrical resistance drops rapidly at the same point as the individual nanoparticles melt and merge to provide a continuous conducting path.

Apart from the thermal properties of the target nanoparticles, the optical properties of the material are critical for laser sintering, since photothermal heating characteristics are directly related to the optical absorption of the nanoparticles. Particles at nanoscale, noble metallic nanoparticles in particular, exhibit unique and tunable optical properties due to their surface plasmonic resonance, and it has been extensively studied for numerous applications. By tuning the nanoparticle size and morphology, together with strong absorption at Rayleigh scattering regime, efficient and local energy deposition can be achieved by a laser. Since laser sintering is commonly applied to nanoparticle film, optical properties of a nanoparticle film are often more practically meaningful in the laser sintering process. Pan et al. [10] extracted the optical properties of silver nanoparticles from spectroscopic ellipsometry and determined its refractive index (n, k) and thickness from direct fitting of measured data. It was shown that a single Lorentz oscillator well explains the dielectric function of the as-deposited film, while two oscillators are required for the sintered films. Such tunability in the optical property is a huge benefit in laser processing as it allows strong absorption at a specific wavelength (**Figure 1(d)**) to minimize possible damage at other non-processed regions [11].

Sintering is often a many-body problem, yet sintering between two individual nanoparticles has been also investigated through both molecular dynamics simulation and experiments. Necking between two gold nanoparticles can occur even at room temperature (300 K) by local potential gradient, and it is insensitive to laser irradiation as the initial growth occurs very fast (<150 ps) [12]. Molecular simulation shows that major neck growth mechanisms during laser sintering can vary according to the particle sizes, which might include grain-boundary sliding/dislocations, surface diffusion, and viscous flow [12]. Laser-induced nanowelding between two metallic nanoparticles has been experimentally confirmed with TEM. Kim et al. [13] reported that gold nanospheres at 13-nm diameter were successfully welded with picosecond laser pulses with 30-ps duration. TEM grids loaded with gold nanoparticles were directly irradiated with the laser, and its high-resolution TEM (HRTEM) image approved that the joint nanoparticles had single-phase nanocontact to have ohmic electrical connection. The time scale required for the absorbed photon energy to be converted into heat is known to be in the picosecond regime [14], however, the nanoparticle coalescence time and dynamics were not well explained earlier. Pan et al. [15] studied laser-induced coalescence of gold nanoparticles supported on a quartz substrate for the investigation of the coalescence time. The gold nanoparticles were fabricated using e-beam lithography for a more systematic study, and pump-and-probe technique has been utilized with 527-nm pulsed laser and 633-nm continuous laser as the processing laser and probing laser, respectively. Time-resolved transmission traces suggested that the coalescence time for melted nanoparticles is at nanosecond scale as shown in **Figure 2(a)**, and this value is reasonably in agreement to the molecular simulation result.

Practically, single laser sintering involves a large number of nanoparticles, and the overall process should be examined in a more macroscopic perspective. Paeng et al. [16] suggested that silver nanoparticle film under continuous wave-laser irradiation goes through certain steps of initial contact, neck growth, and coalescence for complete sintering, and the macroscopic coalescence time measured from pump-and-probe technique is around 10 ms. This value is in accordance with the time lag between laser irradiation and conductive metal electrode formation that was measured by transient resistance change during the laser irradiation

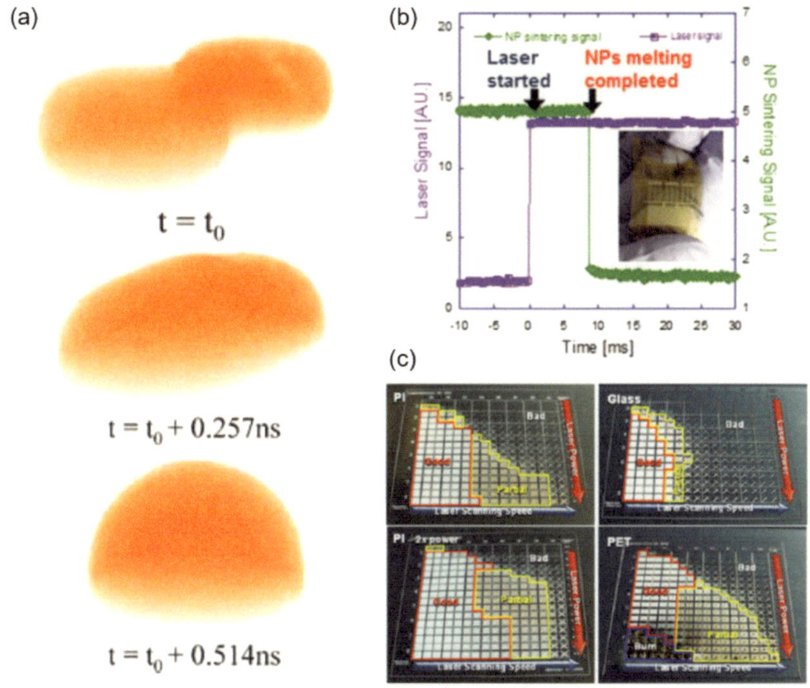

Figure 2. (a) Snapshots of the dynamic coalescence process between two gold nanoparticles from molecular dynamic simulation. Reprinted with permission from Ref. [15]. Copyrights 2008 American Institute of Physics. (b) Transient resistance change measurement according to the laser irradiation. (c) Parametric study for optimum laser conditions on various substrates. Reprinted with permission from Ref. [9].

on silver nanoparticle ink as shown in **Figure 2(b)** [9]. Since overall laser sintering process is a complex multi-physics problem, which incorporates a number of unexpected issues such as balling defects [17], a parametric study over laser power and scanning speed is often conducted to find the optimum sintering condition experimentally [9]. A typical example of the parametric study is shown in **Figure 2(c)** with silver nanoparticle film on various substrates. Ag nanoparticle ink was coated on three different substrates—polyimide (PI), polyethylene terephthalate (PET), and glass—and a parametric study in terms of laser power and laser scanning speed was conducted. It is apparent that the status of sintered Ag nanoparticle can be distinguished into three categories, and the area subject to higher laser power and lower scanning speed shows more complete sintering features. Note that there exists damage threshold for plastic substrates such as PI and PET. Due to the monochromaticity of lasers, wavelength is not an easily adjustable parameter in many cases, but the effect of laser wavelength in the sintering process has also been investigated for silver nanoparticles by utilizing three different (diode) lasers [18]. Paeng et al. utilized three laser diodes at 405, 514.5, and 817 nm, whose optical penetration depths are 28.3, 43.5, and 171 nm, respectively, for the Ag nanoparticle film. It was found that surface melting morphologies were observed for the laser wavelength with short optical penetration depth due to the strong surface absorption. The threshold laser powers for sintering were different according to the laser wavelength as well.

3. Materials

In the early stage, metal nanoparticles, the noble metals in particular, were the main target materials for laser sintering due to their low electrical resistance and superior chemical stability. Gold [19–27] and silver [4, 8, 9, 28–35] are the most widely studied materials for laser sintering as conductors at microscale. These nanoparticles are usually prepared in the solution form with a specific solvent at high weight percentage, while small amounts of surfactants are added to prevent unwanted agglomeration or enhance the dispersion of nanoparticles within the solution. For the efficient use of laser, Bieri et al. [19] controlled the diameter of gold nanoparticles so that the absorption depth is minimized at the laser wavelength as confirmed from effective medium theory, Rayleigh scattering, and Mie scattering. Once a focus laser scans the gold nanoparticle film for selective sintering, the remaining nanoparticle solution is washed away with the same solvent used for the nanoparticle ink. It is frequently reported that the topography of the resultant conductor changes significantly with laser parameters [21, 22]. The most obvious trend is that the linewidth widens as the laser power becomes larger, since the area subject to higher intensity than the threshold value for the initiation of the sintering process increases. The morphology of laser-sintered gold nanoparticle line measured by atomic force microscopy (AFM) often shows a bowl-shaped geometry which can be attributed to thermocapillary effects that arise from the Gaussian profile of a focused laser beam. This trend can be also found from the sintering of other metal nanoparticles as well [29]. At high incident laser power, the cross-section morphology of the sintered gold line becomes "sombrero"-like with the rough surface topography [21], and the possible reason behind such a phenomenon might be the substrate deformation as the maximum temperature at the center region exceeds the softening temperature of the underlying substrate. These morphologies are shown in **Figure 3(a)**. Besides examining the resultant morphology of the sintered metal line, macroscopic electrical conductivity is an important factor to evaluate the performance of the laser as a sintering tool. The resultant electrical conductivity is dependent on a number of factors such as particle size, irradiated laser power, and translation speed [24], but it can be as low as only two times higher (5.41 μΩ cm) than the bulk value (2.65 μΩ cm) as in **Figure 3(b)**. The difference could be explained by boundary scattering from polycrystalline structures and trapped residual capping agent inside the sintered conductor. The laser power density for the strong coalescence of gold nanoparticles was measured to be in the range of 9000–14,000 W/cm^2 [21].

Silver nanoparticle is a good substitute for gold nanoparticles as it shows comparable electrical conductivity with high chemical stability at relatively economical price. The resultant minimum resistivity of laser-sintered silver nanoparticle was as low as 1.59 μΩ cm [9], which is only 130% of the resistivity of its bulk counterpart. One of huge benefits of laser sintering is that the entire process is conducted in ambient conditions at low temperatures which allows using heat-sensitive, cheap polymer as a substrate. A combinatorial study has been thoroughly done for silver nanoparticle to find the optimum laser condition for various substrates including widely used polymer substrates such as Polyethylene terephthalate (PET) and Polyimide (PI) film. **Figure 3(c)** shows that the resistivity of the silver nanoparticle film on the PET substrate initially drops and increases dramatically according to the increase in applied laser

http://dx.doi.org/10.5772/intechopen.68872

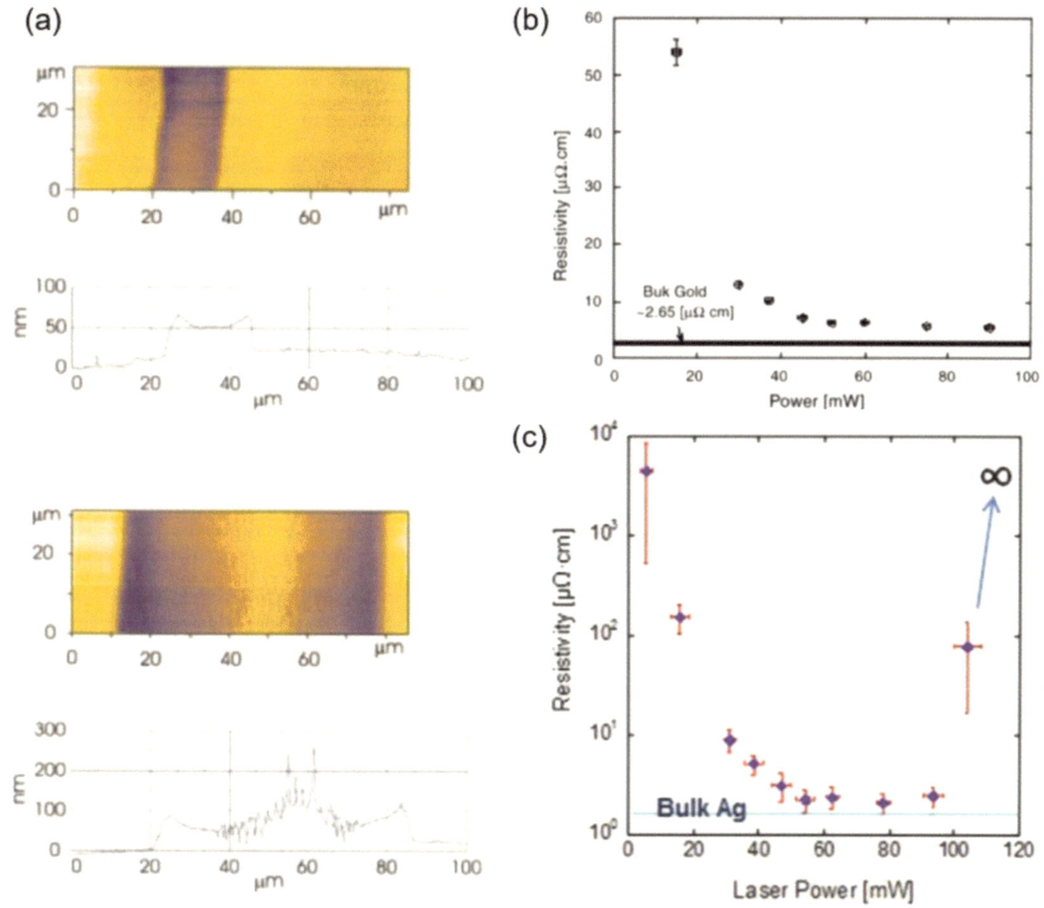

Figure 3. (a) Near-field scanning optical microscopy (NSOM) and atomic force microscopy (AFM) of the sintered gold nanoparticle at (upper) 50-mW and (lower) 300-mW laser power. Reprinted with permission from Ref. [19]. Copyrights 2003 American Institute of Physics. (b) Resistivity of sintered gold nanoparticle electrode at different laser powers. Reprinted with permission from Ref. [24]. Copyrights 2007 Elsevier Inc. (c) Resistivity of sintered silver nanoparticle electrode at different laser powers. Reprinted with permission from Ref. [9].

power due to the thermal damage on the substrate. Minute control of laser power is therefore required for the laser sintering of the metal nanoparticle on flexible substrates. The effect of the underlying substrate has been also confirmed from simulation as well [8]. The mechanical durability of the laser-annealed silver nanoparticle film on the flexible substrate was further tested with outer/inner bending, stretching, cyclic fatigue, and adhesion tests [32]. The laser-sintered silver nanoparticle film shows sufficient mechanical durability, however, void and porosity within the film should be minimized to improve the overall stability.

Recently, more interest has been focused on the laser sintering of copper nanoparticles [36–38] for the replacement of expensive noble nanoparticles. A huge drawback of copper for the sintering process has been its oxidation at an elevated temperature in the ambient condition. Conventional bulk sintering of Cu nanoparticle in ambient condition is often not effective for its application as a conductive layer. Inert gas environment prevents such a problem of Cu nanoparticles during the sintering at high temperature by blocking the oxygen supply required for the oxidation. However, the use of inert gas unavoidably increases the overall production cost as well as the complexity of the manufacturing. Laser sintering provided a possible solution to this problem by reducing the local sintering time. Zenou et al. [36] investigated the resistivity of sintered copper nanoparticle line at different laser power and scan velocities as shown in **Figure 4(a)**. The minimum resistivity value of the laser-sintered copper lines was found to be highly dependent on the sintering time duration, and the lowest resistivity value was achieved at the shortest local sintering time. At high laser scan velocities, the sintered line resistivity almost approached the resistivity obtained under argon atmosphere, indicating that the oxidation problem is effectively prevented even in ambient conditions by reducing the local sintering time. Kwon et al. [37] characterized the effect of laser sintering of the copper nanoparticle in more detail based on diverse analytic tools including scanning electron microscopy (SEM), transmission electron microscopy (TEM), X-ray diffractometer (XRD), Fourier transform infrared spectroscopy (FT-IR), and X-ray photoelectron spectroscopy (XPS) and presented that the laser sintering process offers an enhanced chemical stability together with superior oxide suppression in ambient condition.

The application of laser sintering is not limited to metal nanoparticles, and the other class of target nanoparticles include metal-oxide nanoparticles for their conversion into various functional layers. Metal-oxide often requires different laser conditions; for instance, Pan et al. [39]

(a)

(b)

(c)

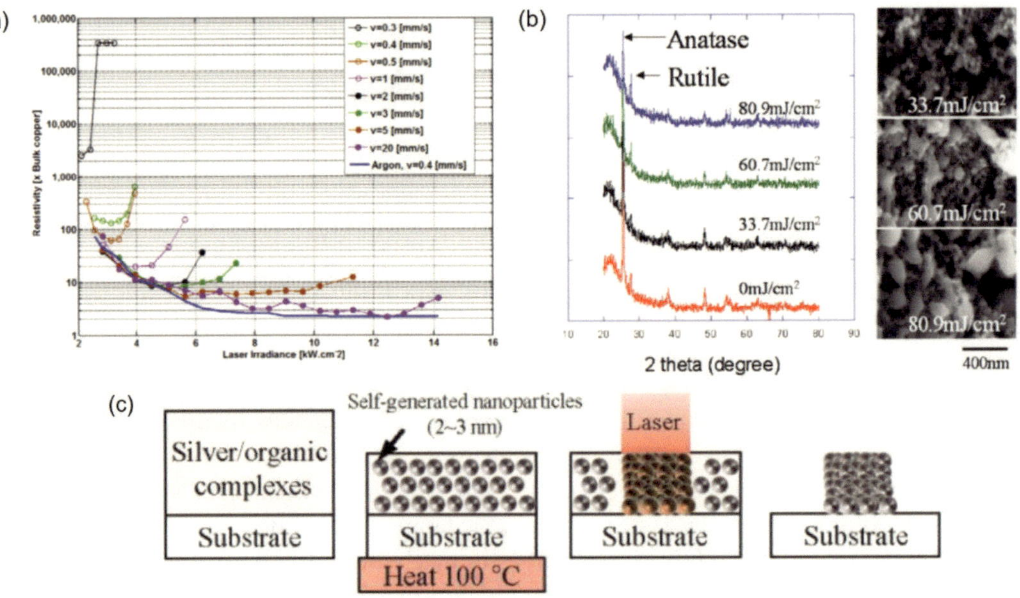

Figure 4. (a) Sintered copper nanoparticle resistivity versus laser irradiance for different laser scan velocities. Reprinted with permission from Ref. [36]. Copyrights 2014 IOP Publishing Ltd. (b) XRD pattern and SEM pictures of TiO2 films on glass after laser annealing with different levels of fluences. Reprinted with permission from Ref. [39]. Copyrights 2012 American Institute of Physics. (c) Process schematics for laser sintering of self-generated nanoparticles from organometallic ink. Reprinted with permission from Ref. [45]. Copyrights 2010 Optical Society of America.

sintered TiO_2 nanoparticles with KrF excimer laser beam at 248-nm wavelength. Laser irradiation not only induced melting and coalescence of these nanoparticles but also phase change from the anatase to rutile for different photocatalytic activity (**Figure 4(b)**). ZnO nanoparticles were also sintered with the same optical configuration, and it was shown that they transform into interconnected porous structures with a single laser pulse of 160 mJ/cm^2 to provide enhanced electrical mobility. For the laser sintering of ZnO [40, 41] or TiO [42] nanoparticles, it was found that the environmental gas largely affects the properties of the resultant sintered film since the surrounding gas changes the amount of oxygen vacancy in the sintered nanoparticle film.

Most of the laser sintering process is solely based on photothermal reactions, but reductive sintering incorporates change in the chemical composition of the material upon the sintering process. Representative examples include reductive sintering of CuO [43] and NiO [44] nanoparticles into copper and nickel films. In the reductive sintering, one of the constituent components such as solvent or capping molecules acts as a reducing agent and converts the metal-oxide nanoparticle into metal. Besides the as-prepared nanoparticles, self-generated nanoparticles from organometallic ink [45], or particle-free reactive ion ink [46] is also subject to the laser sintering process as well, following the procedures as in **Figure 4(c)**.

4. Experimental schemes

Being a direct writing and non-contact method, laser sintering contains a number of strengths including high spatial selectivity and mild environmental requirements, yet various experimental schemes have been developed to solve the remaining weaknesses of the laser sintering process such as inefficient use of material, limited resolution, and throughput. Laser sintering of the nanoparticle is a relatively simple process and typically conducted in three different steps: nanoparticle deposition on the substrate, laser scanning, and removal of the remaining nanoparticle as schematically shown in **Figure 5(a)**. Wet coating process such as spin coating or blade coating is the most direct method to prepare the nanoparticle thin film [9, 29], but a large portion of the nanoparticle is wasted as the remaining nanoparticles are washed away after selective laser sintering process. Various approaches to reuse the nanomaterial are under investigation, and the efforts have been also made in terms of experimental schemes to save the amount of deposited nanoparticles in the first place. Chung et al. [21] integrated a drop-on-demand jetting system in tandem with continuous Gaussian laser irradiation to conduct nanoparticle deposition and sintering in a single platform. As far as jet printing is concerned, background heating was an important factor to evaporate the solvent during or after printing. Followed by jet printing, focused laser was irradiated to the printed line to further melt and coalesce the nanoparticles selectively. Laser sintering of the jet-printed nanoparticle provided both high uniformity and resolution while saving a large portion of wasted nanoparticles. In a similar way, micropipette was also used to write gold nanoparticle ink prior to the laser sintering [20]. Meanwhile, deposition of thin nanoparticle ink is not easily applicable to certain types of substrates such as polydimethylsiloxane (PDMS) substrate. Lee et al. [28] modified the original process and proposed capillary-assisted laser direct writing (CALDW)

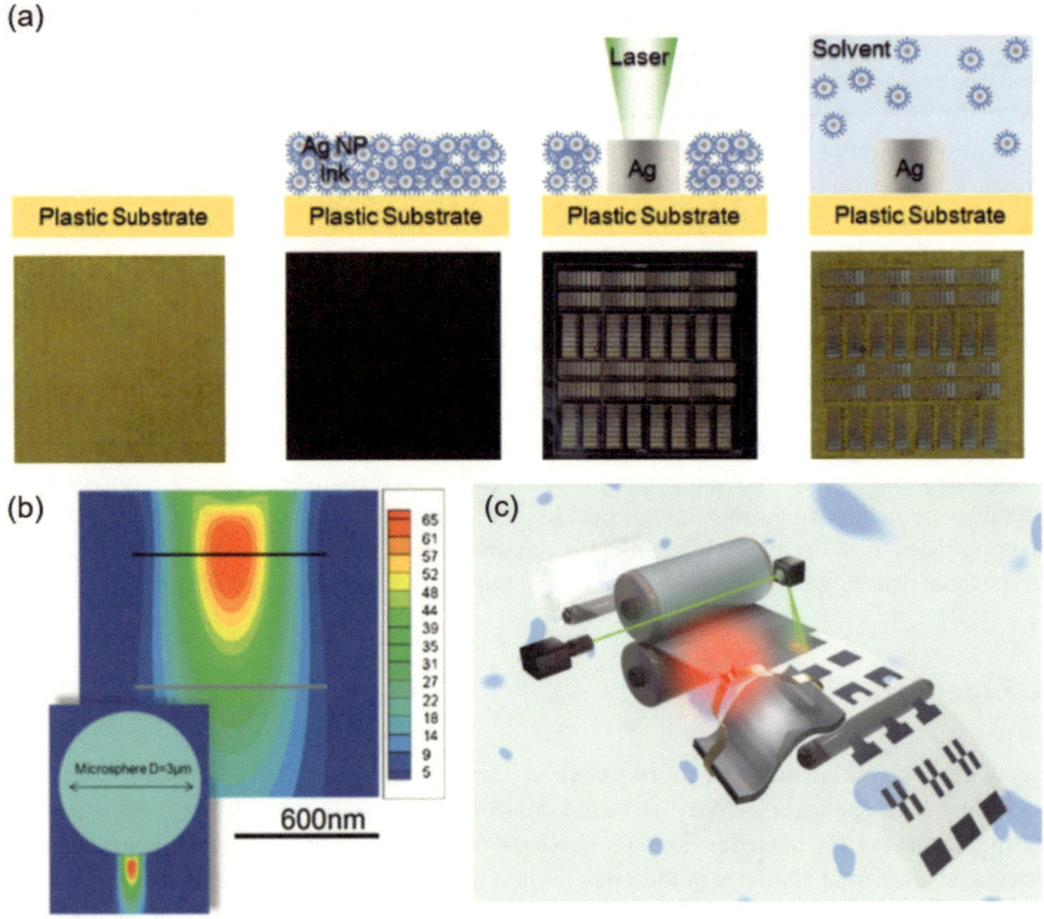

Figure 5. (a) Schematics for typical laser sintering of the nanoparticle. Reprinted with permission from Ref. [9]. (b) Near-field intensity enhancement below the microsphere. Reprinted with permission from Ref. [27]. Copyrights 2010 Wiley-VCH Verlag GmbH & Co. KGaA, Weinheim. (c) A schematic illustration of R2R system integrated with selective laser sintering technique. Reprinted with permission from Ref. [34]. Copyrights 2015 The Royal Society of Chemistry.

of silver nanoparticle for such a substrate that conducts laser sintering of silver nanoparticle in the colloidal environment. On the other hand, some applications require the thickness of the nanoparticle layer to be several microns, which is difficult to anneal with single laser irradiation due to limited optical penetration depth. For the preparation of such a thick nanoparticle film, Pan et al. [39] combined collision nebulizer with laser sintering to achieve simultaneous deposition and sintering in a single configuration.

Compared to other printing techniques, the minimum feature size achievable by laser sintering process is very small—reaching 2-μm feature size by tight focusing [9]—but often not as good as state-of-the-art photolithography process. Sintering is a thermally activated process in principle, and the resultant temperature distribution is largely determined by the area subject to the photothermal heat generation as well as the heat diffusion from the heating spot. The amount of heat generation is proportional to the laser intensity in general, and the spatial distribution of the laser intensity at its focus is determined by two factors in principle—wavelength of the laser and the numerical aperture of the lens used for the focusing, assuming that the optical system is reached at its diffraction limit. In order to push the resolution of laser sintering to nanoscale, ultrashort pulse laser or near-field optics has been employed. Son et al. [4] utilized an ultrashort pulsed laser with pulse duration of less than 100 fs, together with a 100 × oil immersion objective lens, to minimize the focused laser's spot size as well as the thermal diffusion from the heating spot. As a result, the uniform sintered line at 380 nm was fabricated with 780-nm wavelength laser. On the other hand, Pan et al. [27] shrank the focused spot size by using near-field optics. Fluidically assembled microspheres at 3-μm diameter were employed as microlens array as shown in **Figure 5(b)**, and the normalized intensity distribution underneath the microsphere calculated by discrete dipole approximation (DDA) confirming that the full width at half maximum (FWHM) of laser intensity at the designated position can be reduced down to 370 nm. The size of sintered dot diameter was as small as 200 nm, which was further reduced to 50 nm in diameter by the simple post annealing process owing to the densification and beading.

Besides, conventional laser sintering process inevitably suffers from low throughput compared to other mass production techniques due to its direct writing nature, and diverse experimental schemes have been proposed to solve this problem. Instead of moving the sample stage, Yeo et al. [9] scanned the laser beam using 2D galvanometric scanning mirror system to increase the sintering speed to a large extent, up to several meters per second. The laser scanner system was controlled by the CAD software to draw arbitrary 2D images, and it successfully fabricated metal micropatterns on a large polymer substrate over a 4-inch wafer in a fraction of minutes. Pan et al. [42] increased the sintering area per single scanning by using a cylindrical lens that produces the focused spot of elliptical cross-section. A large area sintering was feasible with a relatively small amount of scanning by overlapping the extended focus. On the other hand, An et al. [31] introduced digital micromirror device (DMD) as an optical stamp so that an arbitrary 2D pattern can be sintered at a single exposure without any raster scanning, which is often time consuming. 10 × 10 star-shaped silver electrodes with fine-edge sharpness were fabricated by a step-and-repeat scheme to ensure the potential of this process for the large-area metallic micropattern fabrication. The throughput of laser processing can be further increased as the laser system is compatible to other mass production processes. As a proof of the concept, Yeo et al. [33] integrated the laser sintering scheme with roll-to-roll (R2R) printing system to replace the conventional furnace annealing process and boost the area subject to the laser process. The conceptual image of laser sintering process integrated with the R2R system is illustrated in **Figure 5(c)**.

5. Applications

In the early stage, conducting microlines fabricated by laser sintering of metallic nanoparticles was a good alternative for photolithographically defined metal patterns, especially

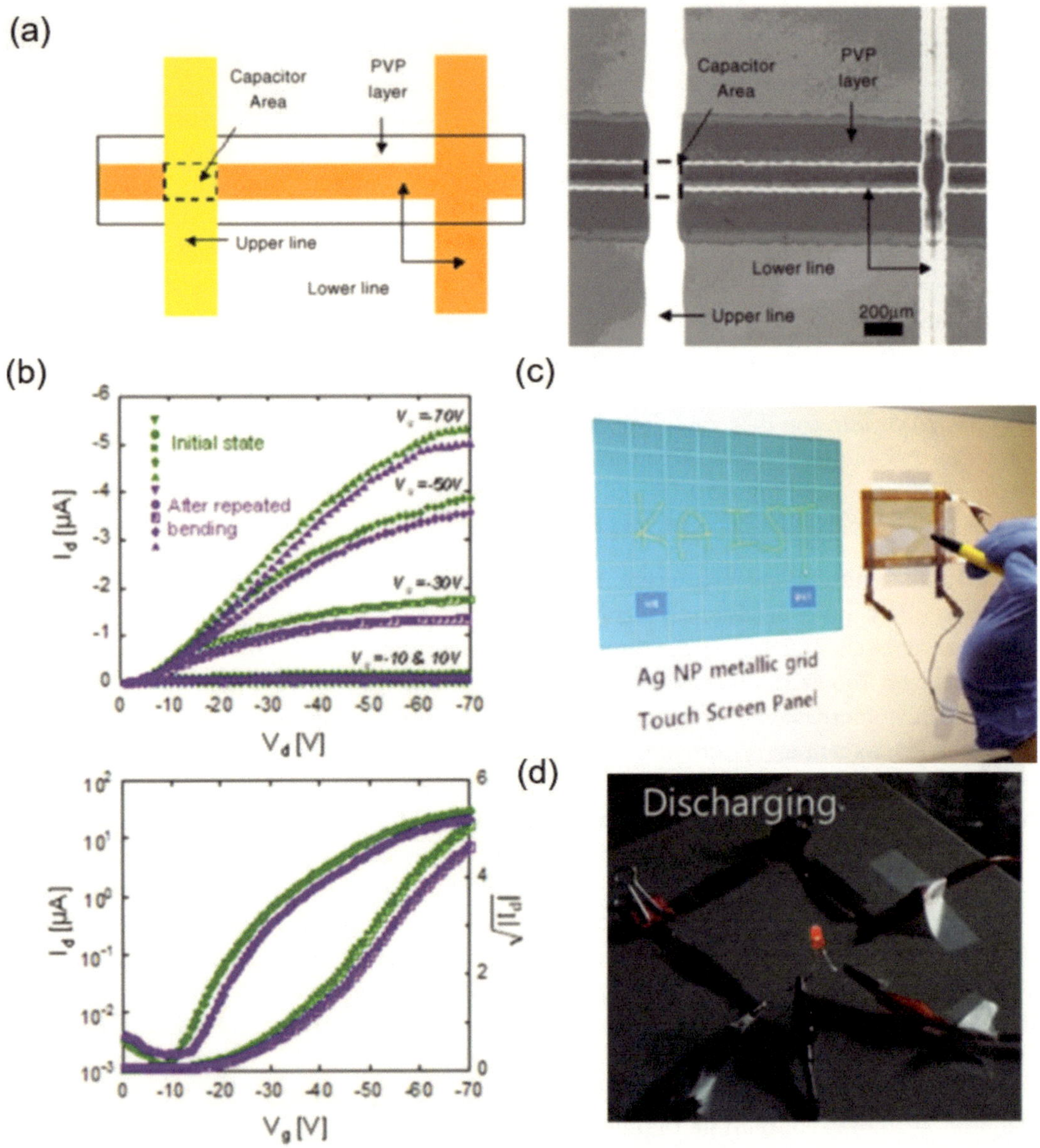

Figure 6. (a) Crossover capacitor schematics on the PI film. Reprinted with permission from Ref. [24]. Copyrights 2007 Elsevier Inc. (b) Output (top graph) and transfer (bottom graph) characteristics of OFET array before and after the bending cycle. Reprinted with permission from Ref. [9]. (c) Demonstration of a metallic grid transparent conductor-based touch panel fabricated by selective laser sintering process. Reprinted with permission from Ref. [29]. Copyrights 2013 American Chemical Society. (d) Demonstration of a series-connected supercapacitor based on laser-sintered nanoparticle current collector to power LED. Reprinted with permission from Ref. [34]. Copyrights 2015 The Royal Society of Chemistry.

for flexible electronics which is not often compatible to the conventional fabrication processes. Laser-sintered microlines were applied to both passive and active electronics, and Ko et al. [24–26] fabricated various electronic components on rigid substrates as well as polymer substrates by applying successive laser sintering for multilayer configuration. The simplest passive electronic component that requires multilayer structure is a crossover capacitor. For its fabrication, a lower level conductor line was firstly ink-jet printed and processed with laser on a PI film, followed by printing of dielectric layer, and another micro-conductor line as shown in **Figure 6(a)**. The measured capacitance for ~200-nm dielectric layer thickness was 1–10 pF for non-shorted capacitors [24]. A transistor is a representative example of active electronic components, and it has been proven that the conductor microlines fabricated by laser sintering of metallic nanoparticles can well replace the electrodes including gate, source, and drain. For the fabrication of the organic field effect transistor (OFET), carboxylated polythiophene with increased air stability was used as the semiconducting polymer. The carrier mobility of the resultant OFET on silicon wafer with laser-sintered source and drain was measured to be around 0.01 cm^2/V, and this value is comparable to the lithographically processed OFET fabricated with the same semiconducting polymer [26]. By expanding the idea and incorporating large-area scanning method, 11,520 OFET arrays were fabricated on a flexible PI substrate with two successive laser sintering processes for source, drain, and gate electrodes [9]. The transistor performance characteristics were measured before and after the 100,000-times bending cycles and confirmed no significant changes in the OFET performance as shown in **Figure 6(b)**. This result validates that the electrical performance of the devices fabricated by the current process is acceptable for broad applications in flexible electronics. On the other hand, Lee et al. [41] utilized laser-sintered ZnO nanoparticle to replace the active channel layer.

Starting from the simple microconductor lines in minute electronic components, potential applications of laser-sintered nanoparticles have been extended to a large extent. Hong et al. [29], Suh et al. [38], and Lee et al. [44] applied laser sintering of silver, copper, and nickel nanoparticles, respectively, to fabricate a large-area flexible transparent conductor based on a metal microgrid structure and utilized it for other optoelectronic devices such as a touch screen panel (**Figure 6(c)**) As the area subject to the laser sintering has increased, laser-sintered nanoparticles find its application in energy devices as well. R2R-printed silver nanoparticles followed by rapid laser annealing is proven to be an efficient way of preparing the current collector for various energy devices including supercapacitors [33, 34] as in **Figure 6(d)**, while sintering of the metal-oxide nanoparticle can be employed for the active electrode layer of solar cells [39]. Besides, the usage of laser-sintered nanoparticles is still under investigation as in the example of the nanocomposite of a nanowire-reinforced nanoparticle matrix film [35].

Acknowledgements

This work was supported by the Human Resources Development of the Korea Institute of Energy Technology Evaluation and Planning (KETEP) grant funded by the Korea government Ministry of Knowledge Economy. (No. KETEP20174010201310)

Author details

Sukjoon Hong

Address all correspondence to: sukjoonhong@hangyang.ac.kr

Department of Mechanical Engineering, Hanyang University, Ansan, Gyeonggi-do, Republic of Korea

References

[1] Cummins G, Desmulliez MPY. Inkjet printing of conductive materials: A review. Circuit World. 2012;**38**(4):193-213

[2] Bäuerle DW. Laser Processing and Chemistry. Springer-Verlag Berlin Heidelberg; 2013. pp. 3-62

[3] Eaton SM, Zhang H, Herman PR, Yoshino F, Shah L, Bovatsek J, Arai AY. Heat accumulation effects in femtosecond laser-written waveguides with variable repetition rate. Optics Express. 2005;**13**(12):4708-4716

[4] Son Y, Yeo J, Moon H, Lim TW, Hong S, Nam KH, Yoo S, Grigoropoulos CP, Yang DY, Ko SH. Nanoscale electronics: Digital fabrication by direct femtosecond laser processing of metal nanoparticles. Advanced Materials. 2011;**23**(28):3176-3181

[5] Ko SH, Park I, Pan H, Grigoropoulos CP, Pisano AP, Luscombe CK, Fréchet JMJ. Direct nanoimprinting of metal nanoparticles for nanoscale electronics fabrication. Nano Letters. 2007;**7**(7):1869-1877

[6] Park S-H, Kim H-S. Flash light sintering of nickel nanoparticles for printed electronics. Thin Solid Films. 2014;**550**:575-581

[7] Dreizin EL, Allen DJ, Glumac NG. Depression of melting point for protective aluminum oxide films. Chemical Physics Letters. 2015;**618**:63-65

[8] Son Y, Yeo J, Ha CW, Lee J, Hong S, Nam KH, Yang D-Y, Ko SH. Application of the specific thermal properties of Ag nanoparticles to high-resolution metal patterning. Thermochimica Acta. 2012;**542**:52-56

[9] Yeo J, Hong S, Lee D, Hotz N, Lee M-T, Grigoropoulos CP, Ko SH. Next generation non-vacuum, maskless, low temperature nanoparticle ink laser digital direct metal patterning for a large area flexible electronics. PLoS ONE. 2012;**7**(8):e42315

[10] Pan H, , Ko SH, Grigoropoulos CP. Thermal sintering of solution-deposited nanoparticle silver ink films characterized by spectroscopic ellipsometry. Applied Physics Letters. 2008;**93**(23):234104

[11] Ko SH, Pan H, Ryu SG, Misra N, Grigoropoulos CP, Park HK. Nanomaterial enabled laser transfer for organic light emitting material direct writing. Applied Physics Letters. 2008;**93**(15):151110

[12] Pan H, Ko SH, Grigoropoulos CP. The solid-state neck growth mechanisms in low energy laser sintering of gold nanoparticles: A molecular dynamics simulation study. Journal of Heat Transfer. 2008;**130**(9):092404-092404-7

[13] Kim SJ, Jang D-J. Laser-induced nanowelding of gold nanoparticles. Applied Physics Letters. 2005;**86**(3):033112

[14] Hodak JH, Martini I, Hartland GV. Spectroscopy and dynamics of nanometer-sized noble metal particles. The Journal of Physical Chemistry B. 1998;**102**(36):6958-6967

[15] Pan H, Ko SH, Grigoropoulos CP. The coalescence of supported gold nanoparticles induced by nanosecond laser irradiation. Applied Physics A. 2008;**90**(2):247-253

[16] Paeng D, Lee D, Grigoropoulos CP. Characteristic time scales of coalescence of silver nanocomposite and nanoparticle films induced by continuous wave laser irradiation. Applied Physics Letters. 2014;**105**(7):073110

[17] Lee Y, Zhang W. In mesoscopic simulation of heat transfer and fluid flow in laser powder bed additive manufacturing. International Solid Free Form Fabrication Symposium, Austin TX, USA; 2015. pp. 1154-1165

[18] Paeng D, Yeo J, Lee D, Moon S-J, Grigoropoulos CP. Laser wavelength effect on laser-induced photo-thermal sintering of silver nanoparticles. Applied Physics A. 2015;**120**(4):1229-1240

[19] Bieri NR, Chung J, Haferl SE, Poulikakos D, Grigoropoulos CP. Microstructuring by printing and laser curing of nanoparticle solutions. Applied Physics Letters. 2003;**82**(20):3529-3531

[20] Choi TY, Poulikakos D, Grigoropoulos CP. Fountain-pen-based laser microstructuring with gold nanoparticle inks. Applied Physics Letters. 2004;**85**(1):13-15

[21] Chung J, Bieri NR, Ko S, Grigoropoulos CP, Poulikakos D. In-tandem deposition and sintering of printed gold nanoparticle inks induced by continuous Gaussian laser irradiation. Applied Physics A. 2004;**79**(4):1259-1261

[22] Chung J, Ko S, Bieri NR, Grigoropoulos CP, Poulikakos D. Conductor microstructures by laser curing of printed gold nanoparticle ink. Applied Physics Letters. 2004;**84**(5):801-803

[23] Chung J, Ko S, Grigoropoulos CP, Bieri NR, Dockendorf C, Poulikakos D. Damage-free low temperature pulsed laser printing of gold nanoinks on polymers. Journal of Heat Transfer. 2005;**127**(7):724-732

[24] Ko SH, Chung J, Pan H, Grigoropoulos CP, Poulikakos D. Fabrication of multilayer passive and active electric components on polymer using inkjet printing and low temperature laser processing. Sensors and Actuators A: Physical. 2007;**134**(1):161-168

[25] Ko SH, Pan H, Grigoropoulos CP, Luscombe CK, Fréchet JM, Poulikakos D. All-inkjet-printed flexible electronics fabrication on a polymer substrate by low-temperature high-resolution selective laser sintering of metal nanoparticles. Nanotechnology. 2007;**18**(34):345202

[26] Ko SH, Pan H, Grigoropoulos CP, Luscombe CK, Fréchet JMJ, Poulikakos D. Air stable high resolution organic transistors by selective laser sintering of ink-jet printed metal nanoparticles. Applied Physics Letters. 2007;**90**(14):141103

[27] Pan H, Hwang DJ, Ko SH, Clem TA, Fréchet JMJ, Bäuerle D, Grigoropoulos CP. High-throughput near-field optical nanoprocessing of solution-deposited nanoparticles. Small. 2010;**6**(16):1812-1821

[28] Lee M-T, Lee D, Sherry A, Grigoropoulos CP. Rapid selective metal patterning on polydimethylsiloxane (PDMS) fabricated by capillarity-assisted laser direct write. Journal of Micromechanics and Microengineering. 2011;**21**(9):095018

[29] Hong S, Yeo J, Kim G, Kim D, Lee H, Kwon J, Lee H, Lee P, Ko SH. Nonvacuum, maskless fabrication of a flexible metal grid transparent conductor by low-temperature selective laser sintering of nanoparticle ink. ACS Nano 2013;**7**(6):5024-5031

[30] Son Y, Yeo J, Ha CW, Hong S, Ko SH, Yang D-Y. Fabrication of submicron-sized metal patterns on a flexible polymer substrate by femtosecond laser sintering of metal nanoparticles. International Journal of Nanomanufacturing 2. 2013;**9**(5-6):468-476

[31] An K, Hong S, Han S, Lee H, Yeo J, Ko SH. Selective sintering of metal nanoparticle ink for maskless fabrication of an electrode micropattern using a spatially modulated laser beam by a digital micromirror device. ACS Applied Materials & Interfaces. 2014;**6**(4):2786-2790

[32] Yang M, Chon M-W, Kim J-H, Lee S-H, Jo J, Yeo J, Ko SH, Choa S-H. Mechanical and environmental durability of roll-to-roll printed silver nanoparticle film using a rapid laser annealing process for flexible electronics. Microelectronics Reliability. 2014;**54**(12):2871-2880

[33] Yeo J, Kim G, Hong S, Kim MS, Kim D, Lee J, Lee HB, Kwon J, Suh YD, Kang HW, Sung HJ, Choi J-H, Hong W-H, Ko JM, Lee S-H, Choa S-H, Ko SH. Flexible supercapacitor fabrication by room temperature rapid laser processing of roll-to-roll printed metal nanoparticle ink for wearable electronics application. Journal of Power Sources. 2014;**246**:562-568

[34] Lee H, Hong S, Kwon J, Suh YD, Lee J, Moon H, Yeo J, Ko SH. All-solid-state flexible supercapacitors by fast laser annealing of printed metal nanoparticle layers. Journal of Materials Chemistry A. 2015;**3**(16):8339-8345

[35] Suh YD, Jung J, Lee H, Yeo J, Hong S, Lee P, Lee D, Ko SH. Nanowire reinforced nanoparticle nanocomposite for highly flexible transparent electrodes: Borrowing ideas from macrocomposites in steel-wire reinforced concrete. Journal of Materials Chemistry C. 2017;**5**(4):791-798

[36] Michael Z, Oleg E, Amir S, Zvi K. Laser sintering of copper nanoparticles. Journal of Physics D: Applied Physics. 2014;**47**(2):025501

[37] Kwon J, Cho H, Eom H, Lee H, Suh YD, Moon H, Shin J, Hong S, Ko SH. Low-temperature oxidation-free selective laser sintering of Cu nanoparticle paste on a polymer substrate for the flexible touch panel applications. ACS Applied Materials & Interfaces. 2016;**8**(18):11575-11582

[38] Suh YD, Kwon J, Lee J, Lee H, Jeong S, Kim D, Cho H, Yeo J, Ko SH. Maskless fabrication of highly robust, flexible transparent Cu conductor by random crack network assisted Cu nanoparticle patterning and laser sintering. Advanced Electronic Materials. 2016;**2**(12)

[39] Pan H, Ko SH, Misra N, Grigoropoulos CP. Laser annealed composite titanium dioxide electrodes for dye-sensitized solar cells on glass and plastics. Applied Physics Letters. 2009;**94**(7):071117

[40] Pan H, Misra N, Ko SH, Grigoropoulos CP, Miller N, Haller EE, Dubon O. Melt-mediated coalescence of solution-deposited ZnO nanoparticles by excimer laser annealing for thin-film transistor fabrication. Applied Physics A. 2009;**94**(1):111-115

[41] Lee D, Pan H, Ko SH, Park HK, Kim E, Grigoropoulos CP. Non-vacuum, single-step conductive transparent ZnO patterning by ultra-short pulsed laser annealing of solution-deposited nanoparticles. Applied Physics A. 2012;**107**(1):161-171

[42] Pan H, Lee D, Ko SH, Grigoropoulos CP, Park HK, Hoult T. Fiber laser annealing of indium-tin-oxide nanoparticles for large area transparent conductive layers and optical film characterization. Applied Physics A. 2011;**104**(1):29-38

[43] Kang B, Han S, Kim J, Ko S, Yang M. One-step fabrication of copper electrode by laser-induced direct local reduction and agglomeration of copper oxide nanoparticle. The Journal of Physical Chemistry C. 2011;**115**(48):23664-23670

[44] Lee D, Paeng D, Park HK, Grigoropoulos CP. Vacuum-free, maskless patterning of Ni electrodes by laser reductive sintering of NiO nanoparticle ink and its application to transparent conductors. ACS Nano. 2014;**8**(10):9807-9814

[45] Kang B, Ko S, Kim J, Yang M. Microelectrode fabrication by laser direct curing of tiny nanoparticle self-generated from organometallic ink. Optics Express. 2011;**19**(3):2573-2579

[46] Liu Y-K, Lee M-T. Laser direct synthesis and patterning of silver nano/microstructures on a polymer substrate. ACS Applied Materials & Interfaces. 2014;**6**(16):14576-14582

Chapter 8

High-Pressure High-Temperature (HPHT) Synthesis of Functional Materials

Wallace Matizamhuka

Additional information is available at the end of the chapter

http://dx.doi.org/10.5772/intechopen.72453

Abstract

High-pressure techniques have been used extensively in effecting phase changes in materials science for decades. The use of high-pressure high temperature enables changes in material atomic arrangement or structure which in turn brings about changes in functional properties such as magnetism, optical, electrical and thermal conductivity. High-pressure technology is highly specialised and requires understanding to fully utilise its potential as a tool for the development of new and novel functional materials with improved properties. This chapter explores the various high-pressure technologies available and how they have been utilised to obtain a wide range of functional ceramic materials for a wide range of applications.

Keywords: high-pressure high-temperature sintering, functional materials, high-pressure synthesis, phase transition

1. Introduction

The discovery of novel properties and quantum states at high pressure has led to a number of new functional material categories. Pressure has long been recognised as a fundamental thermodynamic variable which can be used to manipulate electronic, magnetic, structural and vibrational properties of materials for a wide range of applications. High pressure effectively decreases the atomic volume and increases the electronic density of reactants which results in unusual and interesting properties. There are two basic approaches evident to high-pressure synthesis which involves structural transformation on the one hand and formation of new chemical bonds on the other. Particularly noticeable discoveries in high-pressure physics include metallisation of hydrogen, quantum criticality, high T_c superconductors, polymorphism and exotic metals [1].

The impact of the pressure dimension has not been very conspicuous because a number of substances that exist at high pressure cannot be retained at ambient pressure [2]. Strictly speaking pressure-induced transitions result in *metastable phases* whose properties may or may not change reversibly over a period of an experiment or observation [2]. Such a metastable condition is enabled by the existence of an energy barrier which when surpassed results in a transition to a thermodynamically equilibrium ground state [2]. This explains the crystallisation of amorphous materials upon heating and the conversion of diamond (high-pressure phase) to graphite (ambient pressure phase) at temperatures above 1500 K under normal pressure [2]. Brazhkin in his analysis explains the stability of high-pressure phases on the basis of a simple thermodynamic argument. Simply put, a metastable high-pressure phase possesses Gibb's free energy, G higher than that of a stable phase [2]. He went on to argue that the equilibrium melting temperature of a metastable crystal ($^*T_m$) is always lower than that of the equilibrium melting temperature of the stable phase (T_m). This basically explains why all metastable high-pressure phases transform to more stable phases upon heating at far lower temperatures than the melting temperature of the stable phase (T_m) [2].

The advancement of high-pressure technology has not been as widespread as other synthesis techniques such as high temperature and catalysis [2]. This emanates from the fact that most high-pressure apparatus are quite complex and costly and the volumes of material obtainable at high pressure are very small [2]. Despite these negative factors, the apparatus that have been developed up to date have given rise to some of the most intriguing material properties ever known to mankind. Lower static pressures in the range 0.1–1 GPa are achievable in fairly large volumes ~0.01–1 m^3 using gas containers, piston-cylinder type and autoclave presses [2]. Pressures in excess of 1 GPa can only be achieved in apparatus which are mechanically operated [2].

2. High-pressure technology and apparatus

High-pressure synthesis can be broadly divided into static and dynamic technologies. Traditionally there are two most widely used static-type apparatus, namely, the piston-cylinder and Bridgeman anvil type. The piston-cylinder pressure zone can accommodate volumes in the range 1–1000 cm^3 for pressures of up to 1 GPa [2]. They can reach maximum pressures of 3 GPa where smaller pressure zone volumes and very large presses are needed [2]. The Bridgeman anvil type can achieve very high pressures depending on the type of anvils: for hard alloy anvils, pressures in the range 15–20 GPa, for SiC anvils 20–70 GPa and for diamond anvils 100–300 GPa are achievable [2]. Commercial presses which incorporate concepts from the Bridgeman anvil and piston-cylinder designs have been developed to synthesise superhard materials obtaining pressures over 5 GPa with volumes of ~1 cm^3 [2]. The most widely used presses in this class are known as the *belt*, multi-anvil apparatus and toroid [2]. Another important laboratory-scale high-pressure tool is the diamond anvil cell (DAC). The DAC is used extensively for exploratory high-pressure synthesis and for characterisation of materials under high pressure. It basically consists of two gem-quality diamonds (~1/3 carat) with flat surfaces (culets) capable of compressing small samples on a metal gasket containing a small hole to accommodate the sample. The DAC can reach very high pressures in excess of 150 GPa and can be heated easily using infrared lasers reaching temperatures in excess of 5000 K [3].

Alternative methods which make use of dynamic pressures generated through the shock wave technique allow for pressures of ~100–1000 GPa over short times of dynamic pressure action (nanoseconds) [3]. The shock wave technique has been used to produce materials in large volumes typically 1–10 cm^3 at pressures of 10–100 GPa [2].

2.1. Theory of sintering

Sintering involves powder compaction at elevated temperatures with or without pressure application to obtain densified solid compacts. Thus the driving force of sintering is a function of surface-free energy, temperature and pressure of the system. At low temperatures and pressures (typically <0.5 T_m), the energy available is insufficient to allow for diffusional mass transport in solid-state sintering of most materials with micron-sized particles [4]. The diffusion coefficient, D, is temperature dependent and follows an Arrhenius relationship with the absolute temperature, T as follows:

$$D = D_0 \, exp\left(-\frac{\Delta H}{RT}\right) \tag{1}$$

D_0 is a constant dependent on the atomic planar distance, and the mean frequency of vibration, ΔH, represents the enthalpy change associated with overcoming the diffusion energy barrier, and R is the gas constant.

The application of pressure during sintering is an effective way of improving the rate of densification as it lower diffusion distances between adjacent particles. This is well articulated by the various creep equations which relate creep rate (linear strain rate) to the densification rate (volumetric strain rate). During hot pressing, the density D of the sample increases with a decrease in thickness L. The variables D and L are related as follows:

$$\frac{M}{A} = LD = L_0 D_0 = L_f D_f \tag{2}$$

M represents the sample mass, A the cross-sectional area and the subscripts 0 and f refer to initial and final states. Differentiating Eq. (2) yields

$$L\frac{dD}{dt} + D\frac{dL}{dt} = 0 \tag{3}$$

This can be simplified to

$$-\frac{1}{L}\frac{dL}{dt} = \frac{1}{D}\frac{dD}{dt} \tag{4}$$

This equation relates the linear strain rate of the body *(dL/dt)* to its densification rate *(dD/dt)*. This is the basis of the Nabarro-Herring creep equation which in essence argues that self-diffusion within the crystal will cause the solid to deform in an attempt to relieve stress [5].

The creep being a result of atoms diffusing from interfaces subjected to a compressive stress (where they have a higher chemical potential) towards those subjected to a tensile stress (lower chemical potential). Herring derived an important relationship between the atomic fluxes ε to the grain size, G:

$$\varepsilon'_c = \frac{40}{3}\left(\frac{D_e \Omega P_a}{G^2 kT}\right) \tag{5}$$

where D_e is the lattice diffusion, Ω atomic volume, P_a applied pressure, G grain size, k Boltzmann constant, T absolute temperature and ε'_c creep rate.

The value of ε'_c is the creep rate equivalent to the linear strain rate $(1/L)dL/dt$ in (5) above. It can be deduced that the densification rate can be enhanced by an application of pressure, P_a, and a reduction in grain size G; the grain size reduction is more enhanced due to the higher power factor. The use of ultrafine nanometric powders has been proven to lower sintering temperatures owing to shorter diffusion distances between particles [6]. Existing experimental evidence has proven that faster densification rate allows full density to be attained at smaller grain size before serious grain growth occurs [6]. In principle, the mechanisms that promote densification (increased temperature) are also responsible for grain growth, and both these mechanisms are proportional to the reciprocal of grain size [6]. Thus the control of the two competing mechanisms presents a challenge in the production of fine–/nano-grained microstructures under conventional sintering.

2.2. Synthesis of functional materials under high pressure

As discussed earlier, the high-pressure techniques are capable of tuning the structure and properties of materials resulting in the synthesis of novel materials. The most widely used high-pressure techniques consist in the synthesis of equilibrium high-pressure phases that can be maintained after the release of pressure [2]. The low- and high-pressure phases are associated with thermodynamic stability regions in the pressure and temperature (P, T) diagram of the material [2]. It is noteworthy to mention that in a number of cases under moderate temperatures, pressure transformations result in energy intermediate *kinetic* phases which are associated with lower activation barrier for transformation [2]. Furthermore, high pressure can also modify micro- and macrostructure of a material at nano- and meso-level, i.e. grain size, texture, morphology, defect structure and concentration [2]. That means the morphology and structure of a material can easily be manipulated by varying the P,T conditions. A number of high-pressure materials have shown intriguing characteristics, among them superconducting, semiconducting, optical, electron-transport, thermal and dielectric. Moreover, pressure tuning has proven to be an invaluable tool for obtaining material properties more rapidly and cleanly in comparison to chemical techniques [3]. This is because the entropy changes associated with volumetric transitions are comparatively small which makes pressure much easier to treat than temperature [3]. Pressure studies also provide valuable information on the properties of materials under compression. This information has been used to produce the same materials at ambient pressure by chemical tuning, e.g. doping of host lattice with smaller atoms to induce chemical pressure that produces a high-pressure equivalent with a metastable phase [2, 3].

2.3. Superconducting materials

In July 1908 a Dutch physicist Heike Kamerlingh Onnes discovered the superconductivity of mercury. He observed that the electrical resistance of mercury dropped down to zero under cryogen conditions (helium cooling at 4.2 K). The temperature at which the resistance becomes zero is known as the critical temperature, T_c. Most conventional metallic superconductors were found to possess such behaviour under liquid helium making the technology complex and costly because helium has limited supply and is costly compared to nitrogen, for example. Over the years, the T_c value of superconducting state has been increasing in steps with the discovery of more and more superconducting materials. In 1986, Bedworz and Muller discovered high temperature T_c at around 35 K in the La-Ba-Cu-O system. This initiated a rapid development in the research of ceramic oxide superconductors [7]. The term 'high temperature T_c' is generally used in literature to denote superconductors with a $T_c > 30$ K. In 1987, Wu et al. discovered the $YBa_2Cu_3O_y$ (123 phase) which was superconducting at 93 K which was followed by discovery in 1988 of compounds based on Bismuth ($Bi_2Sr_2Ca_2Cu_3O_x$) and thallium ($TI_2Ba_2Ca_2Cu_3O_x$) with transition temperatures near 110 and 125 K, respectively [8–10]. The most exciting feature of the oxide ceramic superconductors is that superconductivity can be achieved in liquid nitrogen (77 K) which has lower cooling costs than helium [7]. Moreover, the superconductivity of ceramic superconductors can be maintained at considerably high magnetic fields than the conventional metallic superconductors [7].

Another interesting group of superconducting materials is the cubic perovskite and its derivatives. Which have been in existence for over three decades now. They possess a cubic structure and low T_c. Values such as $SrTiO_3$ ($T_c = 0.4$ K), Ba(Pb,Bi)O_3 ($T_c = 12$ K) and $BaPbO_3$ ($T_c = 0.4$ K). In the late 1990s, Cava et al. discovered a high T_c perovskite (Ba,K)BiO_3 ($T_c = 40$ K) [11].

As mentioned earlier, the discovery of La-Ba-Cu-O superconductors triggered the experimentation of a number of copper oxide ceramics with high T_c. The copper oxide ceramics possess layered perovskite structure with alternate stacks of CuO_2 sheets and blocks along the c-axis.

The $YBa_2Cu_3O_{7-\delta}$ (YBCO) possesses an oxygen-deficit orthorhombic perovskite structure [7].

The metal atom ratio in the structure is well defined as 1:2:3 the oxygen stoichiometry varies widely depending on the synthesis condition [7]. The oxygen content is key in regulating the structures and electrical properties (1.5 in Fei). In the stoichiometry Y-123, δ is always positive and can vary from 0.0 to as high as 1.0 [7]. An optimum superconducting property can be obtained by maintaining δ at lower value. The YBCO structure undergoes phase transition during heat treatment from a tetragonal-insulating state ($\delta < 0.4$) to orthorhombic superconducting state ($0.5 < \delta < 1.0$) by adjusting oxygen content.

There are two routes to obtaining oxide superconductors, i.e. doping of parent material through cationic substitution or oxygen nonstoichiometry. The more popular method is doping based on oxygen nonstoichiometry compared to cationic (or anion) substitutions. However oxygen doping mechanisms are not completely understood.

The critical magnetic field (H_c) and the critical current density (J_c) are the other two important factors that are critical for superconductivity to occur [7, 12]. The T_c and H_c are intrinsic properties of the material, while J_c can be varied by the microstructure of the material [7, 12]. It has been observed that the critical value of each parameter (T, H and J) for superconductivity varies with the other two, i.e. the critical current density J_c will decrease with increasing H and T [7, 12].

Superconductors are divided into two categories, type I and type II. Type II superconductors show a more complicated magnetic behaviour but are practically more important [7, 12]. The science behind the operation of these superconductors is found elsewhere [7, 12].

The use of superconducting devices is limited due to the fact that most superconductors must be cooled to low temperatures to be superconducting. A great technological revolution may be triggered by the discovery of a room temperature (RT) superconducting material. It is widely accepted based on scientific literature that the highest T_c at 1 atm is 135 K in the Hg-Ba-Ca-Cu-O system which can be increased to about 160 K through the use of high-pressure synthesis.

2.4. High-pressure synthesis of superconductors

Since the application of high pressure, a number of new superconductors have been discovered which include 22 elemental solids bringing the total number of elemental solid superconductors to 51 [13]. High-pressure studies have provided ideas to enhance T_c values through chemical means under ambient pressure. For instance, the substitution of the smaller ion Y^{3+} for La^{3+} to generate lattice pressure in La-Ba-Cu-O has led to a large T_c enhancement [13]. Strictly speaking, the majority of the high-pressure superconductors entered this state as a result of pressure-induced insulator to metal transition [13]. In his theoretical analysis in 2002, Schilling showed that the T_c decreases under pressure for most known superconductors, sometimes rapidly depending on the (dT_c/dP) value. In most simple metal superconductors, dT_c/dP is negative; this arises predominantly from lattice stiffening with increasing pressure [12–14]. On the other hand, transition metals were found not to follow a universal behaviour, thus reflecting complexity and potency of electronic properties in the d-system [13].

High pressure is key in stabilising high-order oxide superconductors and can enhance the T_c to values >100 K [15]. The HP synthesis of oxide superconductors is usually conducted at pressures of up to 8 GPa and temperatures up to 1400°C [15]. The energy developed by pressure in solid synthesis processes is small compared to temperature; moreover, the observed kinetic effects are more profound for high-pressure synthesis [15]. For instance, it takes 70 h to obtain a $LaFeO_3$ perovskite structure at normal pressure and temperature of 1000°C and only 5 min using pressures close to 5 GPa at a constant temperature of 1000°C using Fe_2O_3 and La_2O_3 as precursors [15].

Since the discovery of high-T_c superconducting in the cuprate oxide La-Ba-Cu-O at T_c = 35 K three decades ago, there is still controversy on the mechanism responsible for the superconducting pairing [13]. In 1993, the Hg-bearing cuprates have been reported to possess one of the highest T_c values reaching 130 K for $HgBa_2Ca_2Cu_3O_{8+x}$ (Hg-1223) at ambient pressure [12, 13, 16–18]. These compounds have been synthesised under high pressure (2–4 GPa) to produce suites of Hg-1234 and Hg-1223 in the general formula $HgBa_2Ca_3Cu_4O_{10+x}$ [19]. In a separate study, the T_c value of Hg-1223 was found to increase to 164 K when measured in situ under high pressure (30 GPa) [20]. Despite the complexity of these compounds, researchers have reached consensus on the factors that enhance the value of T_c in the superconducting oxides [13]:

i. Cationic substitution can be varied until an optimal value is reached beyond which the hole carrier concentration in the CuO_2 planes reduces.

ii. T_c is enhanced by increasing the number of CuO_2 planes which lie close together in the oxide structure while maintaining optimal doping.

iii. Defects should be positioned as far from CuO_2 planes as possible.

iv. Develop structures with CuO_2 planes that are as flat as possible.

It is interesting to note that the most studied superconducting property under pressure is the transition temperature T_c which depends on a number of factors such as the system studied, doping level n, type and mobility of defects and sometimes the pressure medium used [13, 21].

High-pressure synthesis has also been used to induce superconductivity in compounds containing rare-earth (RE) elements particularly those containing Pr^{3+} [17]. The parent compound $PrBa_2Cu_3O_7$ (Pr123) is not superconducting under ambient pressure; however, these materials were found to be superconducting with T_c = 52 K under highly oxidising conditions using high-pressure synthesis [17]. The T_c value improved to 97 K under reducing conditions [17]. The high-pressure route offers additional control over experimental parameters which can be used to search for new high T_c materials.

In 2001, magnesium boride (MgB_2), a quasi-2D material with strong covalent bonding within the boron layers, was discovered [13]. The superconductive MgB_2 is produced under high-pressure conditions and possesses a T_c of 39 K. Latbalestier et al. in their work observed the absence of the problematic weak-link behaviour of the high T_c oxides and the ease of synthesis of MgB_2. This basically raised interest for the discovery of high-temperature analogues of MgB_2. Studies carried out soon after the discovery of MgB_2 reveal that T_c decreased under high pressure. A variation in the pressure dependences (dT_c/dP) of the various samples reported were attributed to differences in the samples or shear stress effects in the frozen or solid pressure media [13].

Magnesium diboride can be synthesised under ambient or elevated pressures [22]. However the superconducting characteristics such as critical current density and irreversible magnetic field are very sensitive to material density, impurity content and structural defects. Serquis et al. reported that the reason for the limited current density j_c at high field in MgB_2 could be a result of weak connectivity between domains and the presence of impurities in the grain boundaries of MgB_2 (**Table 1**) [22].

Sample composition	Synthesis conditions	T_c value	References
$HgBa_2Ca_3Cu_4O_{10+x}$ (Hg-1223)	Static high-pressure synthesis: 30 GPa	164 K	[20]
$PrBa_2Cu_3O_2$ (Pr123)	Static high-pressure synthesis	52 and 97 K (under reducing conditions)	[17]
MgB_2	Static high-pressure synthesis.	39 K	—
$LaFeO_3$	5 GPa, 5 min static high pressure	—	[15]

Table 1. A representation of some superconducting materials that have been produced under high-pressure conditions.

2.5. Thermoelectric materials

Thomas Seebeck's accidental discovery in the 1820s proved that a junction between a bimetallic couple (any two metals) generates a potential as a function of the temperature gradient at that junction. This effect was explained later in the form of an equation below:

$$V = S(T_h - T_c) \tag{6}$$

where V is the applied potential, S is a proportionality constant relating the temperature gradient to the potential for the specific bimetallic couple and T_c and T_h are the cold and hot temperatures forming the thermal gradient.

In 1838, Lenz demonstrated that a bimetallic couple was capable of serving a heat sink or source when an appropriate current was passed through it [23]. This concept was coined into an expression referred to as the *Figure of Merit* in 1910 by Altenkitch and was later developed into its modern form by Ioffe in 1949:

$$Z = \frac{S^2 \delta}{\lambda} \tag{7}$$

S represents the Seebeck coefficient, δ is the electrical conductivity, and λ is the thermal conductivity.

The units of Z are inverse of temperature, and it is a common practice to express the figure of merit as a dimensionless value ZT as follows:

$$ZT = \frac{S^2 \delta T}{\lambda} \tag{8}$$

The dimensionless figure of merit ZT is the primary tool for comparison of thermoelectric materials. It ranges from zero for poor thermoelectrics to 1.5 for high-performance thermoelectrics [24].

The magnitude of the Seebeck coefficient gives a measure of the ability of a material to develop an electrical potential in response to an applied thermal gradient. Simply put, the charge carriers in a given material possess kinetic energy which is directly proportional to the temperature. The charge carriers on the hot side possess higher kinetic energy than those on the cold side and will drift towards the cooler side establishing a potential gradient in response to the thermal gradient. Thus to improve the efficiency of thermoelectric generators, it is a common practice to pair an electron conductor (n-type component) with a hole conductor (p-type component) in the thermoelectric circuit. Generally speaking, the Seebeck coefficient is expected to be high to ensure a large steady current, and the thermal conductivity and resistivity must both be low. This essentially explains why thermoelectric research is focused on semiconductors rather than metals or insulators.

Today, the *Figure of Merit* concept is used to develop thermoelectric devices which convert thermal gradients into useful electric current or convert electric current into a heat source/sink. They can be used in systems where power is in demand and waste heat is in ample supply or conversely where power is readily available and temperature control is critical. However, the wide application of thermoelectric devices has been limited due to their inefficiency (~ 3%) compared to other power sources [25]. The main attraction to the use of thermoelectric devices is that they do not require any refuelling, high reliability, quiet operation, moving parts and low environmental impact [26]. Thus they are often used under extreme environments where access is difficult and reliability is mandatory. The thermocouple remains the most common application of the thermoelectric effect providing an important service to temperature control systems. However, recent improvements in thermoelectric materials will likely push power applications such as thermoelectric generators and heaters into the mainstream. Thermoelectrics (TE) are expected to drive environmentally friendly power systems which produce no pollutants, are compact and are available in a wide temperature range [26].

The most advanced thermoelectric materials are Bi_2Te_3- and PbTe-based tellurides [27]. Bi_2Te_3 is used as a refrigerator and possesses a maximum ZT value at 300–400 K, whereas PbTe alloys are mainly used in power generation with a maximum ZT at 600–900 K [27]. It must be noted that there are a few thermoelectric materials reported to possess high performance between 400 and 600 K [27]. However, the widespread application of Bi_2Te_3 and PbTe has been hindered by the toxicity and rarity of these elements [28]. Another group of TE materials is the skutterudite compounds which possess a general formula MX_3 or VM_4X_{12} where M = *Co, Rh and Ir and* X = *P, As and Sb*, and the V represents a vacancy inside the relatively large cages formed by M and X ions [29]. The binary compounds $CoSb_3$ are of interest for applications owing to the large Seebeck coefficient (high power factor $S^2\delta$) and high hole mobility [30]. However, the lattice thermal conductivity λ is very high which limits the TE applications of binary skutterudites. The large vacancies in skutterudite structure form cages into which guest atoms can be introduced without structural distortion of the parent lattice. These vacancies can be partially occupied by rare-earth (RE) ions such as La, Ce and Yb or alkali metal ions such as Ca and Ba [30]. A number of studies have hypothesised that the ions inserted into these cages often rattle around disrupting the phonon modes in the system and reducing the thermal conductivity [25, 30–38]. Furthermore, theoretical calculations in referenced literature reveal that the filling fraction limit (FFL) of RE atoms in $CoSb_3$ depends on the charge state of the atom [29]. RE atoms with a charge state of +2, for instance, exhibit relatively high FFLs in comparison to those of a higher charge state of say +3 [29].

A practical challenge associated with the fabrication of TE materials with a high *ZT* is the interdependence of *S*, *δ* and *λ* values. The improvement in one parameter usually adversely influences the others [32]. In 1995, Slack proposed a concept based on 'glass-like' thermal conductivity values referred to as the phonon-glass electron crystal (PGEC) [31]. A number of approaches have been adopted to improve the TE performance of skutterudites such a void filling and lattice atom substitution [32–35]. It is noteworthy to mention that in comparison to n-type $CoSb_3$-based skutterudites which are well researched, the development of p-type skutterudites still lags behind [32, 35]. Elemental filling is effective in suppressing the thermal conductivity *λ* due to the rattling filler atoms (near unit filling fraction), and a low resistivity is ensured by the high hole concentration (>10^{21} cm^{-3}) while maintaining a moderate Seebeck coefficient, *S* [39]. The enhancement of *ZT* by elemental filing is possible through enhancing power factor *PF* ($= S^2/\rho$) and supressing thermal conductivity (**Table 2**).

2.6. Development of thermoelectric materials using high-pressure techniques

The high-pressure technique is one of the modern synthesis methods used to improve the efficiency of TE materials. As mentioned earlier, high pressure enables the synthesis of compounds with a crystal structure or composition which is not achieved at ambient pressure [39]. Bi_2Te_3 and its alloys have been fabricated using a variety of methods which include powder metallurgy techniques such as hot pressing, spark plasma sintering (SPS) methods [40], Bridgman and zone melting techniques [40] and ultra-high-pressure sintering (HPS) methods [40]. High-pressure sintering (HPS) is advantageous in that it provides a low-cost route, is suitable for large-scale production and results in more homogeneous nanocrystalline grain and is effective in restraining grain coarsening during sintering [40].

Recent studies have shown that grain refinement of Bi_2Te_3-based alloys can enhance the thermoelectric performances [40]. In one experiment, a p-type Bi_2Te_3-based nanomaterial was fabricated using HPS, and a ZT of 1.16 was obtained at room temperature (RT) [27]. The nano-grain structure was attributed to effective reduction in the thermal conductivity. Zou et al. obtained an n-type $Bi_2Te_{2.7}Se_{0.3}$ compound doped with Gd through HPS at 6.6 GPa followed by annealing. A maximum ZT of 0.74 was obtained at 423 K which was attributed to the nano-grain

Composition	Synthesis conditions	ZT value	References
p-Type Bi_2Te_3 (nano)	Static high-pressure synthesis	1.16 at room temperature	[27]
n-Type $Bi_2Te_{2.7}Se_{0.3}$ doped with Gd (nano)	Static high pressure at 6.6 GPa followed by annealing	0.74 maximum at 423 K	[40]
Te doped with 0.1% Bi	Static high-pressure synthesis	0.72 at 517 K	[37]
Yb-doped $Yb_{0.29}Co_4Sb_{12}$ (highest Yb doping reported)	Static high pressure at 2 GPa	—	[25]
Ba-filled $Ba_{0.51}Co_4Sb_{14}$ (improved filling for Ba from 0.28 to a maximum 0.51)	Static high-pressure synthesis	0.95 at 883 K	[41]

Table 2. A representation of high-pressure-synthesised thermoelectric materials from selected literature.

structure [40]. HPS also allows high dopant solubility and affects formation energy of defects both of which can strongly influence electrical transport properties [37, 41]. Yang et al. proved this concept by doping Te with 0.1 mol% Bi under HPS to produce a ZT of 0.72 at 517 K which is comparable to that of more complex composition alloys such as Bi_2Te_3 and PbTe [37].

The concept of elemental filling of $CoSb_3$-based skutterudites is most promising in the production of thermoelectric materials working in the mid-temperature range (600–900 K) [41]. The concept involves filling the Sb_{12}-icosahedron voids of $CoSb_3$ crystal which in turn donate electrons to improve the power factor and promote rattling inside crystal voids to scatter phonons and reduce lattice thermal conductivity, leading to enhanced thermoelectric performance [41]. Recently $CoSb_3$-based skutterudites with ZT values as high as 2 have been produced through multiple elemental filling [41]. $CoSb_3$-based skutterudites are traditionally prepared through a process melting, quenching, annealing and final consolidation at ambient pressure [41]. This process usually takes several hours; moreover, theoretical investigations suggested the fillable elements, and their filling fraction limits (FFLs) are restricted using such processing methods [41]. This points to the necessity of utilising alternative preparation methods which can broaden the fillable elements and increase FFL [41]. The high-pressure (HP) synthesis method is effective in lowering the reaction temperature and can promote increased filling fraction into $CoSb_3$ voids [41]. Kang et al. used HP synthesis method to fabricate a Ba-filled skutterudite compound ($Ba_{0.51}Co_4Sb_{12}$) with a maximal ZT of 0.99 reached at 883 K. The high-pressure synthesis was effective in increasing the filling fraction of Ba into the $CoSb_3$ structure from 0.28 to a maximal of 0.51. In a separate study, the highest doping ratio of Yb was reported in the compound $Yb_{029}Co_4Sb_{12}$ synthesised under a pressure of 2 GPa [25]. The FFL value is the highest reported so far and shows a substantial decrease in conductivity with increasing Yb filling ratio x.

Magnesium sulphide (Mg_2S) has been identified as an environmentally friendly TE material likely to replace the TE compounds containing rare elements such as Bi, Te, Pb, Co and Sb in the mid-temperature range (600–900 K) application. Thermodynamic studies have shown that it is difficult to synthesise MgS_2 with a stoichiometric ratio at high temperatures without unreacted Mg [29]. This is attributed to the Mg boiling point (1363 K) which is very close to the melting point of Mg_2S (1358 K) [29]. The use of high-pressure synthesis can control the melting and boiling point temperatures to obtain a stoichiometric Mg_2S ratio without residual Mg. The high pressure was also attributed to a decrease in synthetic temperature which is favourable for the relaxation of the n-type defects resulting in improved TE properties [29].

2.7. Magnetic materials

A wide range of major families of permanent magnets have been in use over the years which include the low-cost and low energy ferrites and the more expensive and higher-performance rare-earth magnetic materials. Alnico an alloy of aluminium, nickel, and cobalt is one of the first magnetic materials developed in the 1930s for military electronic applications. Alnico magnets are known for their high induction levels with good resistance to demagnetisation and stability; they also possess a high working temperature limit (550°C) at a reasonable cost. It is well suited for automotive and aircraft sensor applications. However in comparison to newer materials, Alnico possesses lower coercivity which limits its applications. Ferrite magnets referred to as ceramic magnets were commercialised in the mid-1950s and are the least expensive permanent magnets available. The ferrites are produced through sintering of fine iron oxide mixed with either strontium (Sr) or barium (Ba) and a ceramic binder (MM 2018). Ferrites find use in motors in the form of arc-shaped magnets, in magnetic chucks and magnetic tools. Because of their brittle nature, ferrites are not suitable for structural applications, and moreover their thermal stability is limited to 300°C.

Samarium cobalt (Sm-Co) is the first commercially viable rare-earth permanent magnetic material formulated in the 1970s. Their excellent thermal stability, high corrosion resistance and resistance to demagnetisation make them suitable for high-performance application such as most demanding motor applications and medical applications. It is the most expensive magnetic material on a 'per kg' basis; however, this is offset by the low volume required to fulfil a certain task because of its high energy 16MGOe up to 33MGOe. The most powerful commercial permanent magnets are sintered neodymium-iron-boron (Nd-Fe-B) rare-earth magnets with maximum energy product ranging from 26MGOe to 52MGOe. Nd-Fe-B magnets were developed in the 1950s and have found use in applications where higher efficiency and more compact devices are demanded. They are however prone to oxidation and can only be used at temperatures ≤200°C.

The function of permanent magnetic materials in electric machines is to provide magnetic flux [42]. The most important factors required to achieve this function are the saturation magnetisation ($J_s = \mu_0 M_s$) which is required to be as high as possible and affordable; the other factor is high coercivity, H [42]. There exists two coercivity parameters used to grade magnetic hardness, i.e. intrinsic coercivity, ${}_iH_c$ (or ${}_jH_c$), and technical (or normal) coercivity, H_c [42]. Coercivity is basically the ability to resist demagnetisation either from electric or magnetic circuit and thermal demagnetisation from the operating temperature [42]. Soft magnetic materials possess a typical ${}_iH_c < 1$ kAm^{-1}, and hard magnetic materials have ${}_iH_c$ up to approx. 2800 kAm^{-1} (about 35 kOe) [42]. Permanent magnets suitable for high-temperature applications under strong electric and magnetic circuits must possess high coercivity, ideally ${}_iH_c > H_c$. There are only three permanent magnets which show such a characteristic, i.e. hard ferrites and Nd-Fe-B and Sm-Co magnets [42]. The research focus is the development of Dy-free Nd-Fe-B magnets with high H_c for high-temperature applications [42]. Dy is a very rare metal, and 97% of the world supply is of Chinese origin [42].

In recent years, there is much interest in high-performance permanent magnets based on rare-earth and 3d transition metals (3-TMs) intermetallic compounds [42]. This has been triggered by the need for maximised energy densities at various operating temperatures and to replace the more expensive Dy (dysprosium) in high-performance magnets. Manganese-based magnetic compounds such as Mn-Bi, Mn-Al-C and Mn-Li-N have attracted considerable attention as alternative permanent magnetic materials without critical elements [43]. However, the reaction between alkali earth metals and 3d transition metals to form intermetallic compounds at ambient pressure is not feasible owing to the large difference in atomic radii and electronegativity [2, 43]. Another approach which has been adopted is to incorporate nanoscale soft magnetic phases into a hard magnetic phase matrix to enhance energy density of the composite due to interphase exchange coupling at a reduced cost [42].

2.8. Development of magnetic materials using high-pressure techniques

The use of high-pressure synthesis in magnetic materials has not been fully explored, and there are very limited reports on permanent magnetic alloys consisting of alkali metals (AMs) and 3d metals [43]. With the current research focus towards finding alternatives to traditional high-performance rare-earth magnetic materials, high pressure presents a rare opportunity to be exploited in this area. A majority of materials have been exploited in these groups, the so-called 'multiferroic' (MF) materials which possess ferroelectricity, ferromagnetism and ferroelasticity in a single material [44]. The most desirable property in these materials is the magneto-electric (ME) coupling which is extremely rare in most compounds [44]. Magneto-electric coupling refers to either induction of magnetisation by an electric field or polarisation by a magnetic field [44]. The scarcity of these materials can be explained from the symmetry and electronic properties point of view details of which is discussed elsewhere [44]. Recent studies have proven that ferromagnetism and ferroelectricity can occur simultaneously in a single phase through an additional electronic or structural driving force which supports ME coupling [44]. High-pressure high-temperature synthesis allows stabilisation of metastable or highly distorted structures which might achieve ME coupling [44]. The majority of MF materials exhibit a perovskite structure (ABO_3), typically consisting of corner sharing BO_6 octahedra with B ions (usually magnetic ions such as Mn or Fe) in the centre of the octahedral site and A ions at the centre of a cube formed by eight BO_6 octahedra [44]. The application of HP has been used to induce MF in perovskite compounds by introducing magnetic ions (ferromagnetism) in a non-centromagnetic structure (distorted). Typical examples include $BiFeO_3$ which require high-pressure synthesis to obtain a high-temperature solid-state reaction when using oxides as precursors (Bi_2O_3 melts at 824°C at ambient pressure) [44]. One of the challenges in the synthesis of $BiFeO_3$ (BFO) is the evaporation of Bi_2O_3 at high temperatures during solid-state synthesis also under applied fields of about 200 $kVcm^{-1}$; BFO can decompose leading to hematite Fe_2O_3 [45]. Furthermore, $BiFeO_3$ is metastable in air and above 675°C; it decomposes into various products, and below 675°C the density of the sintered product produced by conventional sintering methods is very low which makes evaluation of electrical and magnetic properties impossible [45].

$BiMnO_3$ possesses a simple perovskite structure and has been obtained under HP/HT synthesis (P = 4 GPa,T = 1273 K) [44]. In 2016, Drygas et al. reported an increase in Mn contents (3–5 at%) in (Ge, Mn)N magnetic materials produced by HP/HT synthesis (7 GPa, 1000°C, 10 min). The Mn solubility in hexagonal GaN is limited to 2.4 at% predicted from a calculated Ga-Mn-N phase diagram [46]. The use of nanomaterials has shown an increase in Mn doping at 7 at% for nano-powders and 10 at% for nanowires [46]. This further proves the advantageous impact of HP/HT synthesis on incorporation of a doping element into a foreign matrix.

The binary Mn-Ge system contains intermetallic phases which exhibit antiferromagnetism with relatively low magnetic ordering temperatures [47]. The phases, Mn_5Ge_3 (T_c = 304 K) and $MnGe_4$ (T_c = 340 K), are the most interesting as they are known to exhibit ferromagnetism with T_c above room temperature [47]. It is clear that high-pressure synthesis has been the method of choice in the development of intermetallic compounds which would otherwise not react under ambient pressure conditions owing to differences in atomic radii and electronegativities. Under high-pressure conditions, for instance, alkali earth metals are able acquire properties of

Composition	Synthesis conditions	Target structure	References
$BiMnO_3$	4 GPa, 1273 K under static high-pressure conditions	Perovskite structure with improved magnetic properties	[44]
Higher Mn doping of (Ge, Mn)N	7 GPa, 1000°C, 10 min under static high pressure	Improved doping (3–5 at%) to above the limit of (2.4 at%) resulting in superior magnetic properties	[46]
Mn_5Ge_3, $MnGe_4$	—	Improved reaction kinetics to form intermetallics exhibiting ferromagnetism	[47]

Table 3. Magnetic properties and corresponding high-pressure synthesis conditions of some representative magnetic materials.

transition metals which enables them to easily interact and react with d-metals to form intermetallic compounds (**Table 3**) [2].

3. Outlook

The above examples are evident of the importance the pressure parameter that plays in the tuning of structural, electronic and magnetic properties of functional materials. The commercial production of functional materials under high-pressure conditions is however still severely restricted owing to the small volume of product obtained at high pressures. The major drawback of high-pressure synthesis is the destabilisation after pressure release. However, high-pressure synthesis still remains a powerful research tool in the discovery of novel materials with unique properties which can be recreated through alternative chemical paths. I agree with the sentiments of most high-pressure specialists that high-pressure research is not yet fully appreciated to its true value.

Author details

Wallace Matizamhuka

Address all correspondence to: wallace@vut.ac.za

Vaal University of Technology, Department of Metallurgical Engineering, Vanderbijlpark, South Africa

References

[1] Mao HK, Chen B, Chen J, Li K, Lin JF, Yang W, Zheng H. Recent advances in high-pressure science and technology. Matter and Radiation at Extremes. 2016;**1**(1):59-75. DOI: 10.1016/j.mre.2016.01.005

[2] Brazhkin VV. High-pressure synthesized materials: Treasures and hints. High Pressure Research. 2007;**27**(3):333-351. DOI: 10.1080/08957950701546956

[3] Badding JV. High-pressure synthesis, characterization, and tuning of solid state materials. Annual Review of Materials Science. 1998;**28**(1):631-658. DOI: 10.1146/annurev.matsci.28.1.631

[4] Pagh HLD, McCormic PG, Rouff AL. Mechanical Behaviour of Materials under Pressure, Creep under High Pressure. Amsterdam: Elsevier Publishing; 1970. pp. 355-356

[5] Rahaman MN Ceramic Processing and Sintering. 2nd ed. NY, USA: Marcel Dekkar; 1995. pp. 254-270

[6] Matizamhuka WR. Spark plasma sintering (SPS)-an advanced sintering technique for structural nanocomposite materials. Journal of the Southern African Institute of Mining and Metallurgy. Dec 2016;**116**(12):1171-1180. DOI: 10.17159/2411-9717/2016/v116n12a12

[7] Togano K, Superconductive ceramics. In: Somia S et al. editors. Handbook of Advanced Ceramics. Elsevier; 2003. p. 241-264. DOI: 978-0-12-654640-8

[8] MK W, Ashburn JR, Torng C, Hor PH, Meng RL, Gao L, Huang ZJ, Wang YQ, Chu A. Superconductivity at 93 K in a new mixed-phase Y-Ba-cu-O compound system at ambient pressure. Physical Review Letters. Mar 2, 1987;**58**(9):908. DOI: 10.1103/PhysRevLett.58.908

[9] Maeda H, Tanaka Y, Fukutomi M, Asano T. A new high-T_c oxide superconductor without a rare earth element. Japanese Journal of Applied Physics. Feb 1988;**27**(2A):L209. DOI: 10.1143/JJAP.27.L209/meta

[10] Sheng ZZ, Hermann AM. Bulk superconductivity at 120 K in the Tl–Ca/Ba–Cu–O system. Nature. Mar 10, 1988;**332**(6160):138-139. DOI: 10.1038/332138a0

[11] Cava RJ, Batlogg B, Krajewski JJ, Farrow R, Rupp LW, White AE, Short K, Peck WF, Kometani T. Superconductivity near 30 K without copper: The Ba0.6K0.4BiO_3 perovskite. Nature. Apr 28, 1988;**332**(6167):814-816. DOI: 10.1038/332814a0

[12] Sathish CI. High pressure synthesis and characterization of non-oxide superconductors. PhD thesis. Japan: Hokkaido University; 2013

[13] Schilling JS. What high pressure studies have taught us about high-temperature superconductivity. In: Frontiers of High Pressure Research II: Application of High Pressure to Low-dimensional Novel Electronic Materials. The Netherlands: Springer; 2001. pp. 345-360. DOI: 10.1007/978-94-010-0520-3_26

[14] Eiling A, Schilling JS. Pressure and temperature dependence of electrical resistivity of Pb and Sn from 1-300 K and 0-10 GPa-use as continuous resistive pressure monitor accurate over wide temperature range; superconductivity under pressure in Pb, Sn and In. Journal of Physics F: Metal Physics. Mar 1981;**11**(3):623. DOI: 10.1088/0305-4608/11/3/010/meta.

[15] Demazeau G, Huppertz H, Pottgen R. High-pressure/high-temperature synthesis in materials chemistry. Zeitschrift für Naturforschung B. Jan 1, 2005;**1**(12):1455-1456

[16] Bednorz JG, Müller KA. Possible high T_c superconductivity in the Ba–La–Cu–O system. Zeitschrift für Physik B Condensed Matter. Jun 1, 1986;**64**(2):189-193. DOI: 10.1007/BF01303701

[17] McMillan PF. High pressure synthesis of solids. Current Opinion in Solid State and Materials Science. Apr 30, 1999;**4**(2):171-178. DOI: 10.1016/S1359-0286(99)00013-3

[18] Ciobanu M, Matsumoto K. Recent advances in organic synthesis under high pressure. European Journal of Organic Chemistry. Apr 1, 1997;**1997**(4):623-635. DOI: 10.1002/jlac.199719970404

[19] Lokshin KA, Kuzemskaya IG, Kulikova LF, Antipov EV, Itskevich ES. High pressure synthesis of Hg-1234 and strongly-overdoped Hg-1223 phases. Physica C: Superconductivity. May 15, 1997;**279**(1-2):11-17. DOI: 10.1016/S0921-4534(97)00084-1

[20] Gao L, Xue YY, Chen F, Xiong Q, Meng RL, Ramirez D, Chu CW, Eggert JH, Mao HK. Superconductivity up to 164 K in $HgBa_2Ca_{m-1}Cu_mO_{2m+2+\delta}$ (m = 1, 2, and 3) under quasihydrostatic pressures. Physical Review B. Aug 1, 1994;**50**(6):4260. DOI: 10.1103/PhysRevB.50.4260

[21] Schilling JS, Klotz S. The influence of high pressure on the superconducting and normal properties of high temperature superconductors. Physical Properties of High Temperature Superconductors III. 1992;**3**:59

[22] Prikhna TA, Savchuk YM, Sergienko NV, Moshchil VE, Dub SN, Nagorny PA, Gawalek W, Habisreuther T, Wendt M, Surzhenko AB, Litzkendorf D. Synthesis and sintering of MgB_2 under high pressure. Processing of High Temperature Superconductors. Apr 3, 2002;**140**:365-374

[23] Ioffe AF. Semiconductor Thermoelements and Thermoelectric Cooling. London: Infosearch; 1957

[24] Caillat T, Fleurial JP, Borshchevsky A. Preparation and thermoelectric properties of semiconducting Zn4Sb3. Journal of Physics and Chemistry of Solids. Jul 1, 1997;**58**(7):1119-1125

[25] Staneff GD. High-pressure synthesis of thermoelectric materials. [Doctoral dissertation]. California Institute of Technology. 2005

[26] Sun B et al. Effect of high-temperature and high-pressure processing on the structure and thermoelectric properties of clathrate $Ba_8Ga_{16}Ge_{30}$. Journal of Physical Chemistry C. 2016;**120**:10104-10110. DOI: 10.1021/acs.jpcc.6b02678

[27] Yang M. High-pressure synthesis and thermoelectric performance of tellurium doped with bismuth. Journal of Materials Science. 2017;**52**:10526-10532. DOI: 10.1007/s10853-017-1180-9

[28] Zhang Y et al. Effect of high pressure on thermoelectric performance and electric structure of SnSe via HPHT. Journal of Alloys and Compounds. 2016;**667**:123-129. DOI: 10.1016/j.jallcom.2016.01.158

[29] Sekine C, Mori Y. Development of thermoelectric materials using high-pressure synthesis technique. Japanese Journal of Applied Physics. 2017;**56**:05FA09

[30] Nolas GS, Cohn JL, Slack GA. Effect of partial void filling on the lattice thermal conductivity of skutterudites. Physical Review B. Jul 1, 1998;**58**(1):164. DOI: 10.1103/PhysRevB.58.164.\

[31] Slack GA, Rowe DM. CRC Handbook of Thermoelectrics. Boca Raton, FL: CRC; 1995. p. 407

[32] Liu Y, Li X, Zhang Q, Zhang L, Yu D, Xu B, Tian Y. High pressure synthesis of p-type $Ce_yFe_{4-x}Co_xSb_{12}$ skutterudites. Materials. Mar 31, 2016;**9**(4):257. DOI: 10.3390/ma9040257

[33] Pei YZ, Yang J, Chen LD, Zhang W, Salvador JR, Yang J. Improving thermoelectric performance of caged compounds through light-element filling. Applied Physics Letters. Jul 27, 2009;**95**(4):042101. DOI: 10.1063/1.3182800

[34] Shi X, Yang J, Salvador JR, Chi M, Cho JY, Wang H, Bai S, Yang J, Zhang W, Chen L. Multiple-filled skutterudites: High thermoelectric figure of merit through separately optimizing electrical and thermal transports. Journal of the American Chemical Society. Apr 27, 2011;**133**(20):7837-7846. DOI: 10.1021/ja111199y

[35] Rogl G, Grytsiv A, Rogl P, Peranio N, Bauer E, Zehetbauer M, Eibl O. N-type skutterudites $(R, Ba, Yb)_yCo_4Sb_{12}$ (R = Sr, la, mm, DD, SrMm, SrDD) approaching ZT ≈ 2.0. Acta Materialia. Jan 15, 2014;**63**:30-43. DOI: 10.1016/j.actamat.2013.09.039

[36] Liu R, Yang J, Chen X, Shi X, Chen L, Uher C. P-type skutterudites $R_xM_yFe_3CoSb_{12}$ (R, M = Ba, Ce, Nd, and Yb): Effectiveness of double-filling for the lattice thermal conductivity reduction. Intermetallics. Nov 30, 2011;**19**(11):1747-1751. DOI: 10.1016/j.intermet.2011.06.010

[37] Yang J, Meisner GP, Rawn CJ, Wang H, Chakoumakos BC, Martin J, Nolas GS, Pedersen BL, Stalick JK. Low temperature transport and structural properties of misch-metal-filled skutterudites. Journal of Applied Physics. Oct 15, 2007;**102**(8):083702. DOI: 10.1063/1.2794716

[38] Qiu PF, Yang J, Liu RH, Shi X, Huang XY, Snyder GJ, Zhang W, Chen LD. High-temperature electrical and thermal transport properties of fully filled skutterudites RFe_4Sb_{12} (R = Ca, Sr, Ba, la, Ce, Pr, Nd, Eu, and Yb). Journal of Applied Physics. Mar 15, 2011;**109**(6):063713. DOI: 10.1063/1.3553842

[39] Sales BC, Mandrus D, Chakoumakos BC, Keppens V, Thompson JR. Filled skutterudite antimonides: Electron crystals and phonon glasses. Physical Review B. Dec 15, 1997;**56**(23):15081. DOI: 10.1103/PhysRevB.56.15081

[40] Zou P, Xu G, Wang S, Chen P, Huang F. Effect of high pressure sintering and annealing on microstructure and thermoelectric properties of nanocrystalline $Bi_2Te_{2.7}Se_{0.3}$ doped with Gd. Progress in Natural Science: Materials International. Jun 30, 2014;**24**(3):210-217. DOI: 10.1016/j.pnsc.2014.05.009

[41] Kang Y, Yu F, Chen C, Zhang Q, Sun H, Zhang L, Yu D, Tian Y, Xu B. High pressure synthesis and thermoelectric properties of Ba-filled $CoSb_3$ skutterudites. Journal of Materials Science: Materials in Electronics. Jun 1, 2017;**28**(12):8771-8776. DOI: 10.1007/s10854-017-6603-5

[42] Gutfleisch O, Willard MA, Brück E, Chen CH, Sankar SG, Liu JP. Magnetic materials and devices for the 21st century: Stronger, lighter, and more energy efficient. Advanced Materials. Feb 15, 2011;**23**(7):821-842. DOI: 10.1002/adma.201002180

[43] Matsushita I, Kamegawa A, Sugimoto S. High-pressure synthesis of new magnetic compound in Mn–Li–N system. Materials Transactions. Oct 1, 2016;**57**(10):1832-1836. DOI: 10.2320/matertrans.M2016096

[44] Gilioli E, Ehm L. High pressure and multiferroics materials: A happy marriage. International Union of Crystallography Jouranl. Nov 1, 2014;**1**(6):590-603. DOI: 10.1107/s2052252514020569

[45] Ponzoni C, Cannio M, Rosa R, Chudoba T, Pietrzykowka E, Buscaglia V, Finocchio E, Nanni P, Łojkowski W, Leonelli C. Effect of low-temperature high-pressure sintering on $BiFeO_3$ density, electrical magnetic and structural properties. Phase Transitions. Nov 1, 2013;**86**(11):1104-1114. DOI: 10.1080/01411594.2013.771738

[46] Drygas M, Janik JF, Gosk J, Gierlotka S, Palosz B, Twardowski A. Structural and magnetic properties of ceramics prepared by high-pressure high-temperature sintering of manganese-doped gallium nitride nanopowders. Journal of the European Ceramic Society. Mar 31, 2016;**36**(4):1033-1044. DOI: 10.1016/j.jeurceramsic.2015.10.027

[47] Takizawa H, Yamashita T, Uheda K, Endo T. High-pressure synthesis of ferromagnetic Mn_3Ge with the Cu_3Au-type structure. Journal of Physics: Condensed Matter. Oct 25, 2002;**14**(44):11147. DOI: stacks.iop.org/JPhysCM/14/11147